Synthesis Lectures on Engineers, Technology, & Society

Volume 26

Series Editor

Caroline Baillie, School of Environmental Systems Engineering, The University of Western Australia, Crawley, WA, Australia

The mission of this Lecture series is to foster an understanding for engineers and scientists on the inclusive nature of their profession. The creation and proliferation of technologies needs to be inclusive as it has effects on all of humankind, regardless of national boundaries, socio-economic status, gender, race and ethnicity, or creed. The Lectures will combine expertise in sociology, political economics, philosophy of science, history, engineering, engineering education, participatory research, development studies, sustainability, psychotherapy, policy studies, and epistemology. The Lectures will be relevant to all engineers practicing in all parts of the world.

George Catalano

Earth in Crisis

A Call for a New Engineering Ethic

 Springer

George Catalano
Department of Biomedical Engineering
Binghamton University
Binghamton, NY, USA

ISSN 1933-3633   ISSN 1933-3641 (electronic)
Synthesis Lectures on Engineers, Technology, & Society
ISBN 978-3-031-13316-9   ISBN 978-3-031-13317-6 (eBook)
https://doi.org/10.1007/978-3-031-13317-6

© The Editor(s) (if applicable) and The Author(s), under exclusive license to Springer Nature Switzerland AG 2022

This work is subject to copyright. All rights are solely and exclusively licensed by the Publisher, whether the whole or part of the material is concerned, specifically the rights of translation, reprinting, reuse of illustrations, recitation, broadcasting, reproduction on microfilms or in any other physical way, and transmission or information storage and retrieval, electronic adaptation, computer software, or by similar or dissimilar methodology now known or hereafter developed.

The use of general descriptive names, registered names, trademarks, service marks, etc. in this publication does not imply, even in the absence of a specific statement, that such names are exempt from the relevant protective laws and regulations and therefore free for general use.

The publisher, the authors, and the editors are safe to assume that the advice and information in this book are believed to be true and accurate at the date of publication. Neither the publisher nor the authors or the editors give a warranty, expressed or implied, with respect to the material contained herein or for any errors or omissions that may have been made. The publisher remains neutral with regard to jurisdictional claims in published maps and institutional affiliations.

This Springer imprint is published by the registered company Springer Nature Switzerland AG
The registered company address is: Gewerbestrasse 11, 6330 Cham, Switzerland

This painting needs a reference A Communion of Subjects by Paul Waldau and Kimberly Patton, Columbia University Press, 2009

To the four-leggeds including Brownie, Captain, Snickers, Nikki, Lupino, Francesca, Gianni, Spooky, Isabella, Joey, Luca and especially Yukon who have taught me so very much about life. They have a special place in my heart.

To Brother Wolf in hope that someday humankind will find a safe place in its world for you. As Barry Lopez once wrote, "The gaze of the wolf reaches into our soul."

To Saint Francis of Assisi whose life of service to others, including all of Creation and especially those whose quiet voices are often unheard, has served as a role model for my life.

Preface

> *Here we are, the most clever species ever to have lived. So how is it we can destroy the only planet we have?*
>
> —Jane Goodall

Jane Goodall's question is one that has troubled me for most of my career in engineering. Rather than extending compassion and respect to all the inhabitants of the Earth, too often we view other people, animals, and the natural world as a hindrance—blocking what we deem as progress. As we rush headlong into the future, it seems that our inability to see the interconnectedness of all of life results in our unconscious neglect, misuse, and/or mistreatment of other members of the web of life upon which we all depend.

Engineering has been particularly good to me in many ways. I have had success and have been able to maintain a comfortable lifestyle for my family. I have had textbooks and journal articles published and been awarded numerous research grants from federal agencies on both technical subjects and educational ideas and insights. I have spoken at venues around the world on many of my insights and discoveries. As an educator, I have worked with countless, brilliant young minds, nurturing them along their career paths toward success. I have had the honor of being a member of university faculties both in the USA and in Europe as well as experiencing exciting private consulting opportunities.

Through it all, though, something has seemed wrong. Beginning in the sixties when I was a young engineering student, an uncomfortable sense of foreboding began, and continues to this day, as a deep concern about the ways engineering has responded, or in some cases, failed to respond to a range of contentious issues, chief among them the exploitation and deterioration of the natural world. This book is my effort to not only reflect on the present global environmental crisis from an engineering perspective, but also understand how our profession has arrived at its present relationship to the natural world. Beyond that reflection and understanding, my purpose is to offer a new engineering ethic that will enable us as a profession to take a more inclusive and expansive view of our responsibilities as we work to slow down the rapidly deteriorating health of the Earth.

Dante wrote in the beginning of the Inferno, "In the midst of life's journey I found myself in a dark wood, for the right path was lost." Because I too have felt lost at times as I wrestled with the many difficult challenges facing engineering, it is my hope that this text will help to foster a new conversation with others in the engineering profession who, like me, are troubled by the seeming indifference our profession has displayed toward those with whom we share this beautiful planet.

Binghamton, USA George Catalano

Acknowledgements With gratitude and appreciation for my life's partner, Karen, who inspires me with her love, wisdom, and grace. Without her extraordinary and dedicated editorial help and otherworldly support, this book would never have been completed successfully.

Contents

1 **The Choice** 1
   1.1 Introduction 1
   1.2 Cultural Paradigms 3
   1.3 Moving Beyond the Classical Mechanics Paradigm 4
   1.4 The Choice 4

2 **The Engineering Lens and Classical Mechanics** 7
   2.1 Introduction 7
   2.2 The Mechanical Universe 10
   2.3 Ethical Implications of a Mechanical Universe 13
   2.4 Engineering and Seeing Through the Lens of a Mechanical World 13
   2.5 A New Paradigm Emerges 14

3 **A New Story of the Universe** 17
   3.1 Beginnings of Quantum Mechanics 17
   3.2 Quantum Mechanics Properties and Implications 19

4 **Engineering Ethics in a Mechanical World** 23
   4.1 The Beginnings 23
   4.2 Growing Awareness 24
   4.3 Engineering Ethics Today 27

5 **A New Engineering Ethic in a Quantum World** 31
   5.1 The Intersection and the New Story 31
   5.2 The New Ethic 32
   5.3 A New Engineering Discipline: *Omnium* Engineering 37

6 **The Ethical Implications of the Mechanical and Quantum Perspective: A Case Study** 39
   6.1 History of Wolves in Yellowstone National Park 39
   6.2 Wolf Reintroduction in Yellowstone 42
   6.3 Wolf Management: A Comparison of Old and New Scientific and Ethical Paradigms 44

	6.4 Wolves and the Quantum Universe	46
	6.5 Approaching a Decision	48
7	**Final Reflections**	51
	7.1 Intergovernmental Panel on Climate Change	51
	7.2 Other Issues	52
	7.3 Moving Forward	54
	7.4 The Choice Ahead	55

Appendix ... 57

Bibliography ... 59

List of Figures

Fig. 2.1	Medieval great chain of being	9
Fig. 2.2	Contemporaneous painting of Marseille during the Great Plague in 1720	10
Fig. 2.3	Two slit experiment and origin of quantum mechanics (Lederman et al. 2010)	15
Fig. 4.1	Expanding wave of ethical responsibility in engineering—broadening ethical responsibility in time	28
Fig. 5.1	Image of the earth rising over the moon from Apollo 8. (NASA astronauts captured this powerful image 50 years ago. SCIENCE & SOCIETY PICTURE LIBRARY VIA GETTY IMAGES)	33
Fig. 5.2	Evolving engineering ethic (Catalano 2022)	38
Fig. 6.1	Conceptual diagram showing effects of gray wolf reintroduction into the Greater Yellowstone Ecosystem (Ripple 2014)	41
Fig. 6.2	A pasture filled with spherical cows (Shelton and Cliffe 1999)	46
Fig. 6.3	Trophic cascade (Encyclopædia Britannica)	48
Fig. 7.1	Elements of omnium engineering (Catalano 2014)	56
Fig. A.1	Fractal aspect of omnium engineering (Peitgen et al. 1991)	58

The Choice

1

> *The Universe is a communion of subjects, not a collection of objects.*
>
> Thomas Berry (Evening Thoughts, 2015)

1.1 Introduction

In *National Geographic* (2004), in a prophetic article entitled "Signs from Earth," the editors described a list of warnings that indicated a global environmental crisis was already occurring. In response to their provocative question, "What in the world is going on?" they noted rising carbon dioxide levels and increasing global average temperatures—with the temperatures actually spiking particularly at the higher altitudes. Other warnings included the existence of rising sea levels, melting glaciers, thinning sea ice, permafrost thaws, the increasing occurrence of wildfires, shrinking lakes, collapsing ice shelves, dying coral reefs, increasing deforestation, and more sustained droughts in some regions while increasing precipitation in others. The article also pointed to the increasing loss of natural habitats accompanied by rapidly declining biodiversity, a crisis which many experts refer to as the Sixth Extinction.

That article was published in 2004. In the intervening years, each one of those signs has worsened—pointing to an increasingly rapid and deepening global crisis. I cannot help but wonder: What are we as engineers going to do in the face of such an unprecedented crisis? What is our role? What are our responsibilities? What will guide our actions? What will help us make sense of what is going on?

While we began the twentieth century dazzled with the novelty and wonder of airplanes, automobiles, and radio, the century ended with spaceships, computers, cell phones, and the wireless Internet—all being technologies we commonly take for granted today.

Our personal lives have become highly dependent on the technology that engineers have developed as these innovations have profoundly changed the way we live, including the way we purchase products, communicate, and travel.

However, there are significant drawbacks to this unbounded advancement. By a supreme irony, our technological commitment to "progress" and the betterment of the human condition has had devastating consequences for the basic life systems of the Earth upon which our fundamental existence depends.

In 2015 the eco-theologian Thomas Berry wrote:

> In our times, human cunning has mastered the deep mysteries of the Earth at a level far beyond the capacities of earlier peoples. We can break the mountains apart; we can drain the rivers and flood the valleys. We can turn the most luxuriant forests into throwaway paper products. We can tear apart the great grass cover of the western plains and pour toxic chemicals into the soil and pesticides onto the fields until the soil is dead and blows away in the wind. We can pollute the air with acids, the rivers with sewage, the seas with oil — all this in a kind of intoxication with our power for devastation at an order of magnitude beyond all reckoning. We can invent computers capable of processing ten million calculations per second. And why? To increase the volume and the speed with which we move natural resources through the consumer economy to the junk pile or the waste heap. (Berry, *Dream of the Earth* 2015)

It is essential for us to recognize that while we as engineers are responsible for—and can be proud of—the many advances in modern technology, we share responsibility for many of its grim consequences—chief among them being the deterioration of the natural world.

The engineering profession, similar to other professions, is rooted in an ethical code with slight variations depending on the particular engineering discipline. According to the National Society of Professional Engineers (NSPE 2019), the umbrella professional engineering society, professional engineers must take seriously their responsibility—not just for the quality of the jobs they work on—but for the safety and well-being of the public at large. While this is an excellent beginning point, it is unfortunate that there is rarely a mention of any ethical responsibilities towards the environment or the natural world in any of the ethical codes. In fact, we in engineering tend to refer to the environment and natural world as "natural resources," raw materials to be used, managed, and ultimately controlled to suit our needs or whims at any moment. Why is that? Where did that attitude come from?

Engineering, since the beginnings of the Industrial Revolution, has been based primarily upon both Isaac Newton's classical mechanics and the empirical method first proposed by Francis Bacon. Models in classical mechanics describe the action of a system of forces upon the motion of both macroscopic objects such as projectiles and mechanical devices, as well as astronomical objects such as spacecraft, planets, stars, and galaxies. It has been and continues to be viewed as a system that includes all of the physical laws of Nature that account for the behavior of a world in which all events are determined completely and uniquely by previously existing causes, a philosophical model called determinism.

Although classical mechanics has been foundational in many great technical achievements, it is important to note that classical mechanics clearly views Nature and everything in it as separate, disparate, individual objects—a collection of lifeless things. The classical or Newtonian model likens the Universe to a clock—a perfect machine ticking along with its gears governed by the universal natural laws of physics, resulting in every aspect of the "machine" being predictable.

In 1641, Rene Descartes referred to the Universe as nothing more than a "lump of wax." (Veitch 2020) He argued that animals and, by extension, plants, are purely physical entities, having no mental or spiritual substance. Descartes decided that animals cannot reason, think, feel pain, or suffer, comparing them to machines devoid of any consciousness (Descartes 2015).

For Descartes and advocates of the mechanical world model of the Universe, an evening walk through a nearby forest was nothing more than a passage among separate lifeless objects devoid of the ability to respond, reason, communicate, or feel. These philosophers and natural scientists viewed oak, maple, elm, and birch trees as nothing more than props on a stage (though we know now that they not only communicate with each other and nourish each other but also rally to the defense of their cohorts when set upon by invaders). They would have been unaware of the large variety of bird species that are able to warn their young about nearby predators. Nor could they have imagined that birds with their 'bird-brains' were passing along information to both their young and other species detailing the location of a nearby food source.[1]

1.2 Cultural Paradigms

Viewing "Nature-as-machine" is considered an example of a cultural paradigm, a paradigm being a system of beliefs, ideas, values, and habits that represent a way of thinking about the world, a set of assumptions that form a model that has the potential to shape a culture's perception of reality. Cultural paradigms put a box around or frame how we view every aspect of life. As a result, everything from science to religion to philosophy to engineering is viewed through the lens of the current paradigm. Typically, cultural paradigms arise when a small set of people like philosophers, religious leaders or scientists set forth an idea or a model which ultimately influences large masses of people.

But here is a crucial point. Cultural paradigms can shift and when they do, the changes that occur can affect an entire society. Change only comes though when 'something' catastrophic happens that has a major impact on the masses of people. Examples of

---

[1] The term 'bird brain' captured the prevailing view of the intelligence of birds up until the beginning of this century when scientists published a major paper correcting this wrong with latest information that birds have many brain cells (neurons) and that these cells are capable of cognitive thought comparable to non-human primates (Emery 2016).

such catastrophes are the collapse of the Roman Empire under the onslaught of barbarian invaders, the Black Plague, also known as the Black Death pandemic, which swept through medieval Europe resulting in millions of deaths, and the seemingly endless religious wars that occurred throughout the Middle Ages.

Today we find ourselves at another pivotal moment when life on Earth is under assault. Engineering by and large continues to see Nature and the natural world through the lens of the mechanical world or "Nature as machine" paradigm. The steadfast adherence to this world view continues even though so many of the paradigm's ideas about Nature have been shown to be either completely wrong or limited in application (i.e., incomplete). Because this has profound implications for the ethical foundation of the engineering profession. I passionately believe we need a new paradigm or framework now if we hope to stop and ultimately repair the damage already done.

1.3 Moving Beyond the Classical Mechanics Paradigm

Clearly modern science has moved beyond this outdated paradigm of classical mechanics that has its roots in the sixteenth century. The shift began early in the twentieth century, with the emergence of the work of Max Planck and Albert Einstein's theories of relativity as these pioneers along with others provided the foundation for the new science of quantum mechanics.

As we will explore in the subsequent chapters, the principles of quantum mechanics suggests that rather than the natural world being a collection of lifeless objects, a more accurate description of it may be as a web, a network, or a community (i.e., a communion) of subjects. It is this new science that I believe offers us the possibility of innovative scientific insights which will shape a fresh relationship to Nature and the natural world for our profession. Embracing this new model or what Thomas Berry poetically refers to as the "New Story" will inevitably result in dramatic changes in the roles played by many different professions and institutions as we as a society confront the dire warnings raised nearly twenty years ago.

1.4 The Choice

The choice that lies ahead for us in engineering now as the twenty-first century unfolds is clear. Do we remain steadfast in our conception of the natural world as a collection of lifeless objects, a model passed down to us from the time of Descartes, Bacon, and Newton? Or do we move on as science has moved beyond the models of the seventeenth, eighteenth, and nineteenth centuries with the development of quantum mechanics?

It is unfortunate that, although the implications of a quantum world may present a vastly different sense of ethical responsibilities towards the planet which we share with countless other life forms, much of what this new science offers to us as insights remains relatively unknown and unexplored in engineering at this time.

Bringing those insights into a discussion of engineering ethics is the focus of this text. I believe that this new understanding will enable us to envisage a new cultural paradigm and, as a result, create an engineering ethic that will allow us as a profession to address the present ongoing crises on Earth in a meaningful and creative way.

The Engineering Lens and Classical Mechanics

> *We may regard the present state of the universe as the effect of its past and the cause of its future. An intellect which at any given moment knew all of the forces that animate nature and the mutual positions of the beings that compose it, if this intellect were vast enough to submit the data to analysis, could condense into a single formula the movement of the greatest bodies of the universe and that of the lightest atom; for such an intellect nothing could be uncertain and the future just like the past would be present before its eyes.*
>
> Marquis Pierre Simon de Laplace (Gillispie 1997)

2.1 Introduction

The passage from Laplace is often referred to as "Laplace's Demon" because it focuses upon the idea of determinism, the belief that the past completely and uniquely determines the future. In Laplace's world view, everything is considered to be predetermined: there is no chance, no choice, and no uncertainty.

Determinism captures the very essence of the mechanical world. The philosopher Karl Popper likened determinism to a movie recorded on film (Popper 2002). The movie can be advanced, rewound, and watched countless times while the characters, the setting, the plot, the resolution, and the denouement all remain the same. Although this is only one of many models of the Universe that have been proposed over the course of recorded history, it is this paradigm which has served and continues to serve as the lens through which modern engineering views the natural world.

A brief historical review of the different scientific models that have been proposed in Western history clearly illustrates the fact that scientific paradigms change over time—particularly at pivotal moments of crisis.

- In the fourth century BCE, Aristotle proposed a model of the Universe in which the Sun, Moon, planets, and stars all orbit the Earth inside of concentric spheres. Aristotle believed the Universe was finite in space but existed eternally in time. A contemporary of Aristotle, Exodus Eudoxus of Knidos, was the first to introduce an idea the notion of a solar system, although the Earth was still at the center while planets and other stars remained fixed.
- In the third century BCE, the astronomer Aristarchus of Samos first proposed a model that placed the Sun at the center of the solar system.
- In the second century BCE, Ptolemy extended the work of Aristarchus through his development of an Earth-centric cosmology; that is, it starts by assuming that Earth is stationary and at the center of the universe.
- It was not until the mid-sixteenth century that Nicolaus Copernicus published his work, which positioned the Sun at the center of the Universe, motionless, with Earth and the other planets orbiting around it in circular paths, modified by epicycles, and at uniform speeds.
- In the late sixteenth century, Tycho Brahe proposed a model of the Universe which combined what he saw as the mathematical benefits of the Copernican system with the philosophical and "physical" benefits of the Ptolemaic system.
- In the seventeenth century, Johannes Kepler discovered that the planets all move in elliptical orbits, and as a result developed his three laws, published between 1609 and 1621: the orbit of each of the planets is an ellipse, with the Sun at one focus. (Hawkins 2003)
- Also in the seventeenth century, Galileo Galilei made significant contributions to human understanding of the cosmos as his observations using the newly invented telescope supported the theory that had been put forward by Copernicus (Hawkins 2003).

Accompanying the evolving model of the Universe during the Middle Ages, the cultural paradigm that was most prevalent in Western Europe can best be described and visualized as the "Great Chain of Being." (Fig. 2.1). It is a powerful visual metaphor for a divinely inspired universal hierarchy ranking all forms of higher and lower life. Eco described this paradigm as "a place for everything and everything in its place." (Eco 1978). This yearning for strict order in European society was a consequence of the fall of Rome and the chaos of endless barbarian invasions.

This was also the time of the "Black Death" or "Black Plague," the most devastating pandemic in world history (Fig. 2.2). The Black Death peaked in Europe between

2.1 Introduction

Fig. 2.1 Medieval great chain of being

1348 and 1350 with an estimated one-third of the continent's population ultimately succumbing to the disease. Often simply referred to as "The Plague", the "Black Death" had both immediate and long-term effects on human populations across the world. These included a series of biological, social, economic, political, and religious upheavals which had profound effects on the course of world history, especially European history. Historians estimate that it reduced the total world population from 475 million to between 350 and 375 million. In most parts of Europe, it took nearly 80 years for population sizes to recover, and in some areas more than 150 years. For the people at the time, old models of the Universe no longer seemed sufficient.

One of the lasting consequences of the Black Plague was the belief that the natural world was to be feared and ultimately to be conquered. "Wild" nature was seen as wicked and without value, only having value when it was subdued in service of humankind. This cultural attitude towards the natural world was comfortable with the scientific models soon to be developed in the sixteenth century and beyond.

Near the end of the medieval epoch, a new period emerged which was later referred to ironically as the Age of Enlightenment, or the Age of Reason. Philosophers such as Rene Descartes, Thomas Hobbes, Blaise Pascal, Baruch Spinoza, and John Locke advocated using reason as a means of establishing an authoritative system of aesthetics, ethics, government, and even religion, which would then allow human beings to obtain objective

Fig. 2.2 Contemporaneous painting of Marseille during the Great Plague in 1720

truth about the whole of reality. Emboldened by the discoveries in astronomy and classical mechanics, advocates argued that reason could free humankind from the superstition and religious authoritarianism that had brought suffering and death to millions in religious wars.

Looking back over this history, it is clear to see that not only did the models of the Universe frequently change but also that these changes were accompanied by significant revisions and upheavals of cultural attitudes particularly when society is confronted with crisis. This was true when the Roman Empire fell, when plagues and pestilence swept through medieval cities, and when religious wars seemed unending. And it is true at this dangerous transition point today.

We are currently being confronted by a global crisis—one that threatens the continuance of life on the Earth. A new framework undergirding our response as a profession is needed. It is time for the mechanical Universe model to be challenged, modified, rejected and/or replaced—just as previous models have been. Let us begin then by examining the mechanical Universe and its implications for engineering in more detail.

2.2 The Mechanical Universe

First published in 1687, Newton's *Mathematical Principles of Natural Philosophy* dominated the intellectual landscape of physical science throughout the eighteenth century (Newton 2016). Newton was, without question, the culminating figure in the Scientific Revolution of the sixteenth and seventeenth centuries and the leading advocate of the mechanistic vision of the physical world which was first posited by Descartes (Descartes 2015). It was Descartes who divided the world into conscious thinking substances (human minds) and mechanically arranged substances (the rest of Nature). These metaphysical views have become deeply embedded in Western thought and have conditioned us to view the world through Cartesian spectacles.

2.2 The Mechanical Universe

Table 2.1 Properties implicit in a classical (mechanical) world view

Classical world properties
• Locality: Matter and energy are separate entities
• Separateness; distinct separate parts
• Deterministic; Not random or chaotic
• Direct causality; Direct link between cause and effect
• Linear and nonlinearity
• Inertia-based; Passive matter—external forces
• Analytical; Complex problems divided into set of simpler problems

Within his lifetime Newton saw both the rise and triumph of classical mechanics as well as the widespread acceptance among philosophers and scientists of a mechanistic Universe that operates with mathematical precision and predictable phenomena. Newton's laws—and his theory of a clockwork universe in which God established creation and the cosmos as a perfect "machine" governed by the laws of physics—described matter as inherently passive, only to be moved and controlled by "active principles" or what we refer to today as forces. As a "machine" then the Universe could be partitioned into a collection of distinct and separate parts. The principles that describe a mechanistic Universe are (Table 2.1).

Scientific discoveries occurred during the nineteenth which combined to make a mechanistic point of view of the Universe not merely plausible but seemingly comprehensive. These discoveries resulted in the formulation of conservation laws: the conservation of linear momentum, the conservation of energy and the conservation of mass. Conservation of linear momentum is a major law of physics which states that **the momentum of a system is constant if no external forces are acting on the system**. The law of conservation of energy states that the total energy of an isolated system remains constant; it is said to be conserved over time. The law of conservation of mass states that matter cannot be created nor destroyed in any closed system but can only be converted from one form to another.

The combination of these conservation laws provided a systematic framework that paved the way for new scientific discoveries that fueled rapidly advancing technology. It appeared that all inorganic matter could be analyzed with precise and accurate calculation. The assumption was made that, in any particular system, the amount of matter and energy remained constant.

While the mechanistic model seemed to describe the non-living part of the Universe, a problem arose in its application to the living world. Scientists often found it necessary to introduce in some form of concept such as the "vital inner principle" to account for the biological transformations of adaptation, growth, regeneration, and repair. The "vital inner principle" takes the place of the applied forces in inorganic matter. Ultimately, the need for this principle was eliminated with the theory of natural selection developed by Charles Darwin in *On the Origin of Species* (Darwin 2003). Natural selection proposed that evolution could be entirely explained in terms of environmental conditions, without reference to any internal life force. The "vital inner principle" was no longer needed. As a result, scientists decided that life too could be studied empirically and at last be shown to be governed by the same universal laws as that of the rest of the mechanical Universe. By the end of the nineteenth century, the adoption of the mechanistic model was seen to be complete.

During this historical time, scientific discovery and technological innovation worked hand-in-hand, with scientists discovering and studying new phenomena followed by engineers creating new devices. As the predictability of the heavens was replicated in machines such as steam engines and threshing machines with their interchangeable parts, the concrete application of these principles seemed to corroborate a mechanistic physical model. Inventions that relied on newly harnessed means of utilizing electricity, steel, and petroleum spurred the growth of railways and steamships and transformed everything from farming to manufacturing. The end of nineteenth century was the age of machine tools—tools that make other tools. This period also brought us the assembly line which vastly increased the factory production of goods while also increasing efficiency. Clearly the mechanistic model was correct—or so it seemed at the time.

With the triumph of the mechanical world view and the rapidly advancing technological progress that raised the standard of living for many, it was inevitable that science and scientists assumed the roles once held by the Church and by clerics during the Middle Ages. Reliance on faith was replaced by a reliance on science and reason. Only the "physical world" which is the world that we see around us, that we experience with the five senses of our bodies and is governed by science and the laws of Nature, was thought to exist.

This new understanding of life and the Universe dictated by science led to an ideological belief system that came to be known as "scientific materialism. As defined by the twentieth century philosophers William James and Alfred North Whitehead, scientific materialism is the belief that physical reality, as made available to the natural sciences, is all that is present (Billing 2017). In other words, if it cannot be seen or measured it simply does not exist. The ideology of scientific materialism became dominant in academia during the twentieth century as most scientists began to believe that it was based on established empirical evidence and represented the only rational view or model of the world.

Although it is not based on empirical evidence but is simply a set of assumptions or constructs agreed upon in the sixteenth and seventeenth centuries, it is "scientific materialism" that frames our world view today. Some of the assumptions include:

- An objective world exists independent of the observer

- Objects in the Universe are composed of clumps of matter that are separated from one another in time and space.
- The Universe contains some machines that have learned to think (humans) and the rest (plants and animals) that cannot (reminiscent of Descartes "lump of wax").

2.3 Ethical Implications of a Mechanical Universe

Scientific materialism provided a foundation for a uniquely Western version of progress, one that might be extended as a model for the way things are or should be. Scientific materialism has been applied not only to sanctify industrialization, the division of labor, and the concentration of capital and energy in the hands of the "fittest", but it has also been used to justify colonial imperialism that forcibly dragged "primitive" peoples and whole cultures into the "modern" world of the nineteenth century. Concurrently, this paradigm proved equally devastating for the Earth and all of the natural world.

The predominant Western conception of Nature today is exemplified in - and to no small extent is a consequence of—the philosophy of Descartes, in which, as we have seen, Nature is viewed as something separate and apart, to be transformed and controlled at will (Descartes 2015). Descartes' mechanistic conception of Nature inevitably leads to the view that it is possible in principle to obtain complete proficiency and technical control over the natural world. For Descartes, natural objects were simply lifeless lumps of wax, the perfect exemplification of malleability. We can melt the wax and have it take any particular form we desire. We can shave it down to any particular size. We can use it for a while until its usefulness diminishes for us and then melt it down and start the process all over. In the end, it is simply a lump of wax. It has no intrinsic value. The only value it actually does possess is in meeting our needs, desires, or whims. In scientific materialism, there is little to be concerned about from a moral or ethical perspective. This conception of natural objects as wholly pliable and passive is clearly one which leaves no room for anything like a web of ethical obligations.

2.4 Engineering and Seeing Through the Lens of a Mechanical World

So, there we have it. Most engineering today is still practiced using some aspect of classical mechanics and the conservation laws which came together in the philosophy of scientific materialism.

How can we as a profession ever possibly effectively and creatively overcome issues confronting us today such as global climate change or evidence of the sixth extinction of plant and animal species if a "lump of wax" is what we see when we go for a walk along a Gulf coast shore alive with pods of dolphins or a New England forest filled with fall foliage, wildflowers, and countless bird species? Walking on our beautiful planet with a mechanical world perspective would find us alone in a deafening silence because lumps of wax do not vibrate and thus make no sound. That is not the world in which we live, if we but take the time to listen. Rather than deafening silence, we are greeted with a cacophonous symphony of music from winged parents and their young as night makes way for the day's beginning.

There is something else that we sense when we are immersed in the natural world—its sentience. Aldo Leopold, an American philosopher, and scientist, began his career as an ardent advocate for predator control, arguing that predators such as wolves must be eliminated, likening them to "vermin." He actively participated in the wolf elimination program as a government agent. One day while intending to shoot more wolves, he came face to face with the reality of the natural world's intrinsic aliveness and had an epiphany, a moment of sudden revelation. Leopold later recounted this incident in his book *A Sand County Almanac* (1986) in which he questioned the need to manage and control Nature. Leopold wrote:

> We reached the old wolf in time to watch a fierce green fire dying in her eyes. I realized then, and have known ever since, that there was something new to me in those eyes – something known only to her and to the mountain. I was young then, and full of trigger-itch; I thought that because fewer wolves meant more deer, that no wolves would mean hunters' paradise. But after seeing the green fire die, I sensed that neither the wolf nor the mountain agreed with such a view.

The "green fire" described by Leopold has been and continues to be experienced by more and more people including scientists and engineers, pointing us to the need for a quite different model of the natural world.

2.5 A New Paradigm Emerges

The first questioning of the mechanistic paradigm actually began with Thomas Young's famous double slit experiment (1801). Until the 1800s, light was thought to be made up of particles of light traveling in a straight line, like bullets coming out of a gun. Young's experiment demonstrated that light appeared to act like a wave and indicated that Newton was apparently wrong in his view of light consisting solely of particles. With the beginning of modern physics, about a hundred years later, it was realized that light could, in fact, show behavior characteristic of both waves and particles (Lederman 2010) (Fig. 2.3).

In 1927, Davisson and Germer found that electrons demonstrated the same behavior as light in the double slit experiment. Later, atoms and molecules were shown to behave similarly, again both as particles and as waves. More questions were raised about the

2.5 A New Paradigm Emerges

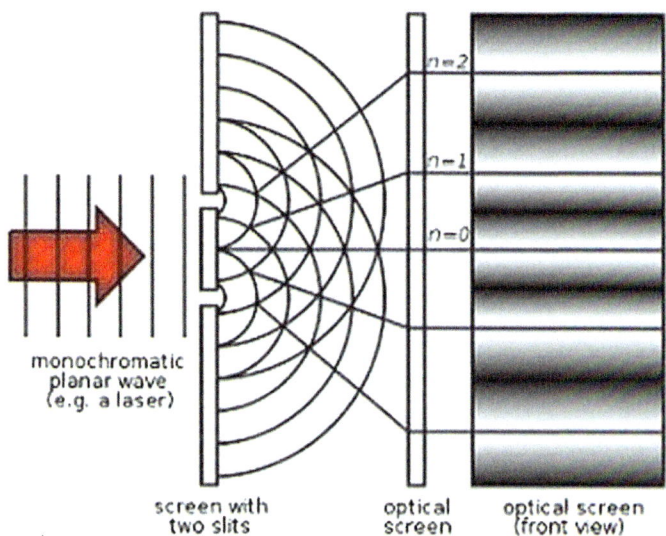

Fig. 2.3 Two slit experiment and origin of quantum mechanics (Lederman et al. 2010)

mechanical world view when Marie and Pierre Curie's experiments with radium challenged the validity of the conservation of force and mass equations. Also, in these early years of the twentieth century, Einstein deduced that time, space, and mass could no longer be considered absolutes.

At the same time, several philosophers and scientists began to question whether the 'laws' of science were, in fact, absolute or immutable. They pondered questions like, might not these 'laws' themselves evolve over time? Might there even be a new model of the Universe? Why shouldn't there be? It had already happened many times in the past.

A New Story of the Universe   3

> *Everything is determined, the beginning as well as the end, by forces over which we have no control…we all dance to a mysterious tune, intoned in the distance by an invisible player. .*
>
> *Albert Einstein* (Einstein 2014)

3.1 Beginnings of Quantum Mechanics

By the end of the 1800s, physicists had discovered empirical phenomena that could not be explained by classical physics. This led to the development, during the 1920s and early 1930s, of a revolutionary new branch of physics called quantum mechanics.

This new science called into question the material foundations of the world by demonstrating that atoms and subatomic particles are not solid objects, nor do they exist with certainty at definite spatial locations and at definite times. Additionally, studies in quantum mechanics revealed that the particle being observed has a direct connection to the mind of the observer; that is to say, both the physicist and the method of observation are intrinsically linked to the phenomenon being observed.

The history of quantum mechanics, which is a fundamental part of the history of modern physics, began essentially with a number of different scientific discoveries and continues development today. Some of the more important dates and events are:

- 1838—Michael Faraday discovered cathode rays.
- 1859–60—Gustav Kirchhoff solved the black-body radiation problem.
- 1877—Ludwig Boltzmann described a theory that the energy states of a physical system could be discrete.

- 1887—Heinrich Hertz discovered the photoelectric effect. In connection with work on radio waves, he observed that, when ultraviolet light shines on two metal electrodes with a voltage applied across them, the light changes the voltage at which sparking takes place.
- 1900—Max Planck proposed a theory that any energy-radiating atomic system can theoretically be divided into a number of discrete "energy elements." (i.e., the individual quantum particles).
- 1905—Albert Einstein further investigated the photoelectric effect reported by Hertz and proved Max Planck's quantum hypothesis that light itself is made of individual quantum particles.
- 1926—Gilbert Lewis termed the quantum particles 'photons.'
- 1935—The thought experiment known as the Einstein-Podolsky-Rosen (EPR) Paradox introduced the concept of entanglement which resulted in a contentious debate concerning the completeness of quantum mechanics as a scientific paradigm.[1]
- 1964—John S. Bell's developed a simple yet elegant theorem which touched upon many of the fundamental philosophical issues that relate to modern physics.[2] In subsequent experiment after experiment, physicists have confirmed that the Bell theorem is correct—with its most important conclusion being that quantum entanglement does take place.

It is important to note the last concept, quantum entanglement, is at the heart of the disparity between classical and quantum physics. This disparity will be discussed further in the next section.

---

[1] The EPR paradox involves two particles that are entangled with each other according to quantum mechanics. Under the Copenhagen (Neils Bohr) interpretation of quantum mechanics, each particle is individually in an uncertain state until it is measured, at which point the state of that particle becomes certain. At that exact same moment, the other particle's state also becomes certain. The reason that this is classified as a paradox is that it seemingly involves communication between the two particles at speeds greater than the speed of light, which is a conflict with Albert Einstein's theory of relativity. The focus of the debate centered on the concept of entanglement and pitted Einstein and David Bohm who believed that ultimately it would be able to develop a predictive mathematical model for all quantum events versus Neils Bohr who held that such a model would not nor ever would exist.

[2] In its simplest form, Bell's theorem states no physical theory of local hidden variables (i.e., predictive model) can ever reproduce all of the predictions of quantum mechanics. The uncertainty in quantum mechanics does not simply represent a lack of knowledge but rather a fundamental lack of definite reality. Until the measurement is made, the particles are in a superposition of all possible states.

3.2 Quantum Mechanics Properties and Implications

Let us identify several of the properties inherent in a quantum world and discuss possible implications for engineering. The first two properties are nonlocality and quantum entanglement.

Nonlocality describes the apparent ability of objects to instantaneously know about each other's state, even when separated by large distances (potentially even billions of light years), almost as if the Universe at large instantaneously arranges its particles in anticipation of future events. Thus, in the quantum world, instantaneous action or transfer of information does appear to be possible. This is in direct contravention of the "principle of locality" (or what Einstein called the "principle of local action"), the idea that distant objects cannot have direct influence on one another, and that an object is directly influenced only by its immediate surroundings, an idea on which almost all of physics is predicated.

Nonlocality suggests that Universe is in fact profoundly different from our habitual understanding of it, and that the "separate" parts of the universe are actually potentially connected in an intimate and immediate way.

Nonlocality occurs due to the phenomenon of entanglement, whereby particles that interact with each other become permanently correlated, or dependent on each other's states and properties, to the extent that they effectively lose their individuality and, in many ways, behave as a single entity. The two concepts of nonlocality and entanglement go very much hand in hand, and, peculiar though they may be, they are facts of quantum systems which have been repeatedly demonstrated in laboratory experiments.

The implications of this property of entanglement are profound. As engineers we are trained to take complex systems, break them down into sets of simpler sub-systems, find a solution for each and then put all the solutions and all the sub-systems back together and arrive at final complete solution for the original system. In a quantum world, this is no longer possible because sub-systems are now understood to no longer be separate; rather they are known to be connected. This connection exists no matter how far apart or how close the sub-systems exist. It was this property, in fact, which was referred to by Einstein as "spooky action at a distance."

The recognition of the connectedness and nonlocality of sub-systems inevitably leads to the question of individuality. What does individuality actually mean? When does it occur? The quantum property of non-separateness reinforces the notion that in quantum mechanics the individuality of systems (i.e., elements) is a virtual impossibility.

While a mechanical Universe is viewed as governed by determinism, a quantum world model states that chance and randomness also exist. This greatly troubled Einstein. In a letter dated December 1926 and addressed to Neils Bohr, Einstein wrote, "The theory [quantum mechanics] produces a good deal but hardly brings us closer to the secret of the Old One [God]. I am at all events convinced that He does not play dice." Notwithstanding Einstein's hope, a quantum world is marked by the absence of certainty.

In place of direct cause and effect as found in a mechanical Universe, a quantum world is marked by both downward and upward causation. The term downward causation was originally described in 1974 by the philosopher and social scientist Donald Campbell as "a causal relationship from higher levels of a system to lower-level parts of that system: for example, mental events acting to cause physical events." (Campbell 1974). Consider the following example of downward causation. Certain psychological states (e.g., prolonged anxiety, embarrassment) can cause physiological effects (heightened blood pressure, eczema, blushing) in a human body. Here we have a mental process (an emotion) leading to a physiological response.

Similarly, upward causation is a phenomenon that is evident everywhere in our Universe as particles combine to form atoms, which form molecules, which form elements and ultimately results in all of life's forms. Small "things" cluster to form larger "things"—from the bottom up (Kim 1984). A visible example in Nature is a hurricane. Hurricanes are extremely strong weather events that remove heat from tropical waters to fuel their fury. These violent storms form over the ocean, often beginning as a tropical wave—a low pressure area that moves through the moisture-rich tropics, possibly enhancing shower and thunderstorm activity. As this weather system moves westward across the tropics, warm ocean air rises into the storm, forming an area of low-pressure underneath. This causes more air to rush in. The air then rises and cools, forming clouds and thunderstorms. Up in the clouds, water condenses and forms droplets, releasing even more heat to power the storm—a bottom-up phenomenon.

In addition, within downward and upward causation, non-linearity, self-organization, and emergence are often also present in a quantum world. Nonlinearity is a term used in statistics to describe a situation where there is not a straight-line or direct relationship between an independent variable and a dependent variable. In a nonlinear relationship, changes in the output do not change in direct proportion to changes in any of the inputs. Emergence suggests the appearance of behavior that could not be anticipated from knowledge of the parts of the system alone. Self-organization implies that there is no external designer controlling the emergent features; rather, the emergent features appear spontaneously.

Here is a summary of the characteristics of a quantum world (Table 3.1).

Perhaps the most significant quantum feature of all is entanglement which provides the foundation for a more holistic and ecological view of Nature. Albeit in different manners and degrees, we dwell in a Universe in which everything is interconnected, and interrelated.

According to Goethe, "In nature we never see anything isolated, but everything in connection with something else which is before it, beside it, under it and over it." (Goethe 2014).

It is my view that quantum mechanics challenges and empowers us to contemplate the Universe from the distinct perspective called for by Goethe. Quantum mechanics describes a Universe which is much richer than the mechanical Universe—one filled with infinite

3.2 Quantum Mechanics Properties and Implications

Table 3.1 Properties in a quantum world

Quantum world properties
• Nonlocality
• Quantum entanglement; non-separateness
• Randomness and chance
• Noncausality; downward and upward causality
• Nonlinearity
• Emergence
• Self-organization

potentials. With all its difficulties and paradoxes, quantum mechanics leads towards a more mature view of Nature, replacing stifling, outdated scientific, philosophical, and theological perspectives.

How might an embracing of the quantum world impact our ethical responsibilities as a profession towards the Earth? Can we articulate a new ethic for engineering to match the new scientific paradigm? I think we can. In the following chapter I shall first briefly outline the present relationship the engineering profession has with the model of the natural world rooted in the ideas of the eighteenth and nineteenth centuries and how it has evolved over the last half century.

Engineering Ethics in a Mechanical World

4

> *Just as ripples spread out when a single pebble is dropped into water, the actions of individuals can have far-reaching effects.*
>
> Dalai Lama XIV

4.1 The Beginnings

While engineering is the application of knowledge in the form of science, mathematics, and empirical evidence to the innovation, design, construction, operation and maintenance of structures, machines, materials, devices, systems, processes, and organizations, engineering ethics is the system of moral principles that apply to the practice of engineering. It establishes and examines the responsibilities held by engineers to society, to their clients, and to the profession. As a scholarly discipline, it is closely related to subjects such as the philosophy of science, the philosophy of engineering, and the ethics of technology.

In the United States engineering grew in status as a recognized and respected profession during the nineteenth century in part due to the introduction of revolutionary technologies which paved the path to modern life. The light bulb extended the day for both work and leisure; the telephone and telegraph allowed for quick communication; safer elevators let cities expand into a new horizon. These advancements and others that occurred during the 1800s shaped the future of public life. Engineering and engineers soon became highly respected throughout society, and, as a result, the profession began to organize itself in order to establish consistent standards of competencies. This effort by engineers was accompanied by the founding of the four original professional engineering societies. They included:

- 1852—American Society of Civil Engineers (ASCE).
- 1871—American Institute of Mining Engineers (AME).
- 1880—American Society of Mechanical Engineers, (ASME); and
- 1884—American Institute of Electrical Engineers (AIEE).

It was not until the twentieth century that the engineering societies actually established formal ethical codes. Up until that time, the concept of an engineer's ethical code was viewed as a personal issue rather than a broad professional concern. This change was in large part due to a series of disasters that shook the public's trust in the engineering profession.

4.2 Growing Awareness

As the nineteenth century drew to a close and the twentieth century began, a series of spectacular engineering disasters occurred that brought public attention to the importance of engineering standards. One particularly horrific accident took place in Ashtabula, Ohio, on December 29, 1876. A bridge over the Ashtabula River collapsed under the weight of a train carrying 160 passengers. After all but the lead locomotive had plunged into the river, the train's oil lanterns and coal-fired heating stoves erupted, setting the wooden railroad cars on fire. Because firefighters saw the accident as out of their authority and made the decision to not get involved, many who survived the initial crash burned to death. Notwithstanding the heroic action of the local town's people who attempted to pull survivors from the wreck, nearly one hundred people were killed. The coroner's report found that the bridge had been improperly designed by the railroad company's engineers, poorly constructed, and inadequately inspected. As a result of the accident, a federal oversight office was established to formally investigate fatal railroad accidents—this serving as a first step towards holding engineers accountable for their work.

A second disaster, called the Boston Molasses Flood, occurred near Boston on January 15, 1919. A large storage tank filled with 2.3 million US gallons of molasses weighing approximately 12,000 tons burst, and a huge wave of molasses rushed through the streets at an estimated 35 mph, killing twenty-one and injuring 150.

These and similar incidents had a profound effect on how the public viewed engineers at the time and forced the profession to confront shortcomings of not only technical and construction practice but also begin to establish ethical standards. One of the profession's responses was the development of formal codes of ethics by three of the four founding engineering societies that explicitly dealt with the impact of technology on humans.

It is important to note that, while these disasters affected how engineers viewed their responsibilities as a profession with respect to the loss of human life, the profession remained silent on issues related to the natural world until other disasters garnered national attention. One disaster which had significant impact on the profession and forced

4.2 Growing Awareness

a reexamination of the engineering profession's relationship to the natural world was the Love Canal tragedy in New York.

Love Canal is an aborted canal project branching off of the Niagara River about four miles south of Niagara Falls. From 1942 to 1953, the Hooker Chemical Company, with government sanction, began using the partially dug canal as a chemical waste dump. At the end of this period, the contents of the canal consisted of approximately 21,000 tons of toxic chemicals, including at least twelve that are known carcinogens (halogenated organics, chlorobenzenes, and dioxin among them). Hooker capped the 16-acre hazardous waste landfill in clay and sold the land to the Niagara Falls School Board, attempting to absolve itself of any future liability by including a warning in the property deed. In 1978, a leaking toxic dump with the chemical waste was discovered buried beneath an elementary school in the working-class residential community of LaSalle in Niagara Falls, New York. The health effects for the residents were staggering, with high incidences of cancer, miscarriages, rare diseases, and birth defects. It marked the first time for a U.S. "state of emergency" to be declared over a human-made disaster and was a sobering lesson about the effects of toxic pollution. The toxic legacies of Love Canal still linger today. Toxins were discovered again in the area in 2011, and residents new to the area continue to report unusual health problems.

Love Canal quickly came to symbolize the looming environmental disaster represented by untold numbers of toxic waste disposal sites scattered throughout America. The engineering profession responded with several of the professional societies beginning to bring attention to the need for careful stewardship of Nature—though it is notable that this growing concern focused solely upon the value of the natural world as a source of resources.

An engineering disaster with much greater loss of life occurred at the Union Carbide India Limited pesticide plant in Bhopal, Madhya Pradesh, India. The Bhopal disaster, also referred to as the Bhopal gas tragedy, resulted from a gas leak accident on the night of December 2, 1984. Due to a combination of operator negligence and critical engineering design flaws, over 500,000 people were exposed to methyl isocyanate gas as the highly toxic substance made its way into and around the small towns located near the plant. It is considered among the world's worst industrial disasters. The official immediate death toll was 2259. Others estimate that 8000 died within two weeks, and another 8000 or more have since died from gas-related diseases. Apart from the human toll, we also cannot ignore the environmental impacts of the disaster. Over 2000 animals were killed by the gas that night, most of them livestock that people relied upon for food. In addition, the heavy gas was absorbed into local rivers, making the water undrinkable and poisoning the fish.

Two years later on April 26th, 1986, the Chernobyl accident occurred. This disaster, considered the worst nuclear disaster in history both in cost and casualties, was a nuclear accident that occurred at the No. 4 reactor in the Chernobyl Nuclear Power Plant, in the north of Ukraine. The initial emergency response, together with later decontamination of the environment, involved more than 500,000 personnel and cost an estimated $68 billion. Again, operator negligence and critical engineering design flaws were identified as the causes. Death tolls remain unknown though estimates range near 10,000. According to the World Health Organization (WHO), approximately **7722 square miles** of land in Europe was contaminated. The exact impact of the radiation depended on whether or not it was raining when contaminated winds passed overhead.

In more recent times, two major oil spills have been responsible for heightening awareness of the impact of the engineering profession upon the natural world. The first, the Exxon Valdez incident, occurred off the coast of Alaska in Prince William Sound, on March 24, 1989. Exxon Valdez, an oil supertanker owned by Exxon Shipping Company bound for Long Beach, California, struck Prince William Sound's Bligh Reef, and spilled 10.8 million US gallons of crude oil over the next few days. Immediate effects include the deaths of between 100,000 and 250,000 seabirds, at least 2800 sea otters, 12 river otters, 300 harbor seals, 247 bald eagles, and 22 orcas, and an unknown number of salmon and herring. Nine years after the disaster, continuing studies documented evidence of continuing negative oil spill effects on marine birds.

A second equally tragic spill occurred in the Gulf of Mexico. The Deepwater Horizon oil spill was an industrial disaster that began April 20th, 2010, in the Gulf of Mexico on the BP-operated Macondo Prospect. The U.S. federal government estimated the total discharge at 4.9 million barrels, making it the largest marine oil spill in the history of the petroleum industry. Most of the impact was on the marine species. Eight U.S. national parks were threatened, and more than 4000 species that live in the Gulf islands and marshlands were affected, including more than 1200 species of fish, 200 species of birds, 1400 species of mollusks, 1500 species of crustaceans, 4 species of sea turtles, and 29 species of marine mammals. After several failed efforts to contain the flow, the well was officially declared sealed on September 19th, 2010, although reports in early 2012 indicated that the well site continued to leak. The Deepwater Horizon oil spill is still regarded as one of the worst environmental disasters in American history.

With the increasing rate of occurrence and strength of hurricanes due to global climate change, the importance that engineering plays in protecting the public has become more apparent. Hurricane Katrina which occurred in late August 2005, was one of the worst disasters in United States history with over 1,800 fatalities and $125 billion in damages. Failures within New Orleans' engineered hurricane protection system (levees and floodwalls) contributed to the severity of the event and have drawn considerable public attention. In the years following Hurricane Katrina, forensic investigations have uncovered a range of issues and problems related to the engineering work.

In addition to these events, rapid advances in engineering and synthetic biology have raised a whole host of new ethical questions for engineering as the line between the living and non-living worlds becomes blurred. Synthetic biology is a multidisciplinary area of research that seeks to not only create new biological parts, devices, and systems, but also to redesign systems that are already found in Nature. The ethical aspects of synthetic biology include three main categories: biosafety, biosecurity, and the creation of new life forms.

As a result of all of these events, engineering societies in recent years have begun to focus more attention upon the importance of the profession's ethical standards, questioning whether these standards are sufficiently expansive. A wide array of engineering and engineering education conferences and journals now highlight ethical case studies, and permanent ethics committee have been established in nearly every professional engineering society. However, it remains the case that the mechanical world view still exists and dominates the mindset of the engineering profession. In mainstream engineering, Nature continues to have no intrinsic value and remains only a collection of objects to be used. The lifeless 'lump of wax' model holds on tenaciously.

4.3 Engineering Ethics Today

Today, according to codes of ethics adopted by various professional engineering societies, members of the engineering profession are expected to exhibit the highest standards of honesty and integrity because it is recognized that the practice of engineering has a direct and vital impact on the quality of life for all people. As stated previously, the first fundamental canon in the National Society of Professional Engineers code of ethics requires that "*…engineers, in the fulfillment of their professional duties, shall hold paramount the safety, health, and welfare of the public.*" (NSPE 2019).

It is important for us as engineers in the twenty-first century to consider what is meant by the term "public"? How is it to be interpreted? Does "public" refer only to the members of society that are linked to the technology at the time that technology is developed? What about future generations? What about the parts of the world that are marked by poverty? What about the environment? What about other species? Does the engineer's ethical code have any concern for the millions of species who inhabit our planet, or is it simply all about one species, our species—humankind?

In fact, engineering ethics has steadily evolved over the last century. Although the actual professional codes have remained relatively static, there is growing awareness and some movement in extending the sphere of ethical responsibility to include an explicit commitment to the health of the natural environment (green engineering), and, to a lesser degree, an effort to focus upon the well-being of people in the parts of the world that endure poverty (humanitarian engineering) and injustice (engineering and social justice).

Metaphorically, the direction of the expanding sense of ethical responsibility from traditional engineering to a more inclusive engineering is analogous to the propagation of waves radiating from the impact of a stone in a still pond. Traditional engineering corresponds to the first wave generated when the impact occurs. This is where the focus has been historically. A second wave moves out from the first wave, which includes concerns about the health of the environment. (Green engineering) That is followed by further propagation both to include a consideration of the poor and impoverished regions of the world (humanitarian engineering) and then to broaden the boundaries even more to include the common good. (Engineering and social justice).

Green Engineering

Green engineering, which encompasses all of the engineering and science disciplines, focuses on the design and synthesis of materials, processes, systems, and devices with the objective of minimizing overall environmental impact (including energy utilization and waste production) throughout the entire life cycle of a product or process.

As can be seen in Fig. 4.1, green engineering extends the boundaries beyond the traditional sense of professional responsibility. According to the Environmental Protection Agency (EPA), green engineering is the design, commercialization, and use of processes and products that minimize pollution, promote sustainability, and protect human health without sacrificing economic viability and efficiency.

Several authors have identified core principles of green engineering. One set of principles which is particularly encompassing has been offered by Anastas and Zimmerman in which they describe Green design principles which use an eco-efficient design methodology, a "cradle-to-grave" methodology which seeks to reduce toxic wastes, reduce energy consumption, reduce the demand on various natural resources, and is subject to rules, regulations, and limitations in its attempts to be what they label as "less bad." (2013).

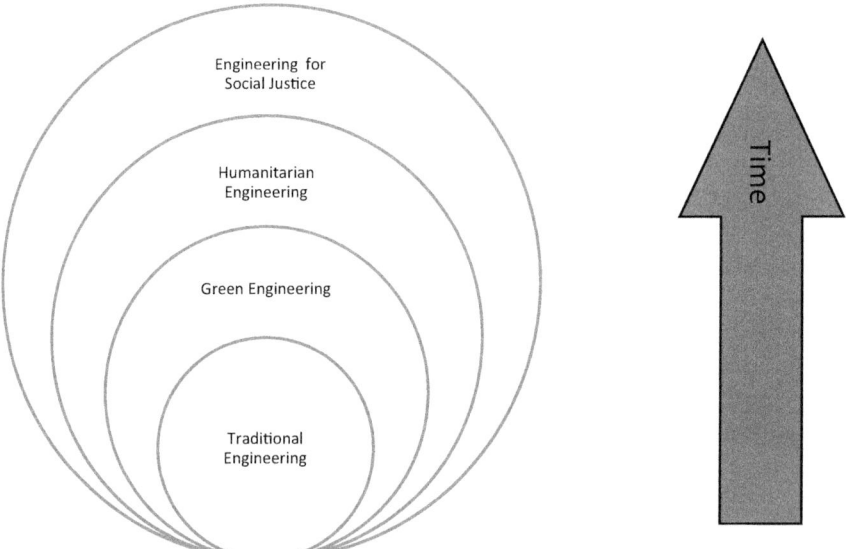

Fig. 4.1 Expanding wave of ethical responsibility in engineering—broadening ethical responsibility in time

Green engineering appears to be growing in acceptance throughout the engineering profession to a certain degree. The NSPE Code of Ethics now encourages engineers to adhere to the principles of sustainable development, defining "sustainable development" as the "challenge of meeting human needs for natural resources, industrial products, energy, food, transportation, shelter, and effective waste management while conserving and protecting environmental quality and the natural resource base essential for future development." While this is a step forward from the traditional engineering approach, the focus still exists on serving the human species alone. The separation remains between humans(us) and the rest of the planet (them—the natural world). Here again Nature appears to only have value if it serves the particular, narrowly focused interests of humans.

Humanitarian Engineering

Humanitarian engineering, a relatively new discipline in engineering, further broadens engineering's sphere of concern. Mitchum and Munoz (Humanitarian Engineering 2010) define humanitarian engineering as "the application of engineering to improving the well-being of marginalized people and disadvantaged communities, usually in the developing world" with an emphasis upon sustainability, low costs, and locally available resources for the solutions advanced. This approach focuses upon local solution, forgoing the hi-tech 'one-size-fits-all' algorithm.

Passino has offered a list of the principles of humanitarian engineering which includes empathy, compassion and building trust and places emphasis upon both developing communities and being aware of the cultural and societal differences that exist (Passino 2016). Passino's principles focusses upon human dignity, human rights, and fulfillment. In humanitarian engineering, the engineering profession has an opportunity to consider the plights of those who traditionally have not been included in engineering classrooms—the poor in this country as well as overseas, as well as the native peoples who were subject to colonialization throughout the Americas.

Humanitarian engineering clearly offers a more expansive view about who or what the "public" should include. It proposes that it is not enough to consider the notion of "public" in a very narrow sense but rather forces an integration of compassion towards all of our fellow human citizens.

Engineering, Social Justice and Peace

At the forefront of the social justice movement is the Engineering, Social Justice, and Peace (ESJP) organization which began at Queens University, Kingston, Ontario in 2004, founded by Caroline Baillie (Baillie, *Engineering and Society: Working Towards Social Justice* 2009). ESJP is committed to envisioning and practicing engineering in ways that extend social justice and peace in the world. This commitment manifests in two major areas: by understanding how technology and society are co-constructed, and by devising and developing technologies and other engineering solutions to the problems that local communities face. An important new element for engineering in this movement is working towards peace.

In summary, while engineering prides itself on being a profession which holds true to its basic canon to "do no harm" to the "public," the breadth and meaning of the word "public" has changed over the course of the last century to include notions of sustainability, humanitarianism, and social justice. That change is to be welcomed. However, what has not changed is the view that we are somehow separate from all that surround us at the most basic level.

We as humans have been conditioned through our education and culture to think of ourselves as the subject, the center point of all of creation. As engineers we have spent countless hours mastering an engineering science that views the entire Universe as a mechanical contrivance that we can take apart, analyze, adjust a few knobs, substitute a few elements, and then put it all back together like a clever watchmaker. The paradigm of scientific materialism assures us that the physical world is all there is. Not much has changed for our profession as we are affixed to that conceptualization from the sixteenth, seventeenth, and eighteenth centuries.

What if we place the mechanical Universe model and scientific materialism aside for a moment and consider the new insights discovered from the science of quantum mechanics? What engineering ethic would then result? How might our relationship to the Earth and all its inhabitants change?

A New Engineering Ethic in a Quantum World

> *A human being is part of a whole, called by us the 'Universe'—a part limited in time and space. He experiences himself, his thoughts, and feelings, as something separated from the rest—a kind of optical delusion of his consciousness. This delusion is a kind of prison for us, restricting us to our personal desires and to affection for a few persons nearest us. Our task must be to free ourselves from this prison by widening our circles of compassion to embrace all living creatures and the whole of nature in its beauty.*
>
> *Albert Einstein (2014)*

5.1 The Intersection and the New Story

While the innovative ideas in quantum mechanics have presented significant challenges in our efforts to understand the natural world, even more challenging is how this resulting new understanding affects our ethical responsibilities towards our profession, society and the rest of the species who live on this Earth. Specifically, the challenge is to find an ethical paradigm that is appropriate for engineering that reflects these new insights.

One approach that holds promise is embodied in the statement written by Thomas Berry who wrote in *The Dream of the Earth*: "The Earth and ultimately the Universe are composed of subjects to be communed with, not a collection of objects to be used." Berry was one of the twentieth century's most prescient and profound thinkers. As a cultural historian, he sought a broader perspective on humanity's relationship to the Earth in order to respond to the ecological and social challenges of our times. He called for a new ethic to capture this new relationship with the Earth and referred to it as the "New Story."

Berry urged humans to recognize their place on a planet with complex ecosystems which exists in a vast evolving universe. He sought to replace the modern alienation from Nature with a sense of intimacy and responsibility. Berry called for not only new forms of ecological education, law, and spirituality but also the creation of resilient agricultural systems, bioregions, and ecocities.

For Berry, the devastation of the planet can be seen as a direct consequence of our separation from the nonhuman world. Historically, this reached its pivotal moment in the seventeenth century proposal of Descartes that the Universe is composed simply of "mind and mechanism." Berry noted that:

> In this single stroke, he (Descartes) devitalized the planet and all its living creatures, except for the human. The thousandfold voices of the natural world [thus] became inaudible to many humans.

Many others have written about the need to respond to the growing ecological crisis we face in the modern world as a result of this separation from the natural world. The psychologist Vaughan-Lee (2016) recognized the heart of the ecological crisis, stating:

> Our present ecological crisis...is calling to us and it is for each of us to respond. This crisis is not a problem to be solved, because the world is not a problem but a living being in a state of dangerous imbalance and deep distress.... There is action to be taken in the outer world, but it must be action that comes from a reconnection; otherwise, we will just be recreating the patterns that have created this imbalance.

The philosopher Douglas Christie (2020) recognized the practice of seeing deeply into the living world as a moral activity that dissolves dualistic thinking and restores a sense of the whole. He described the essential importance of contemplative ecology:

> Our ecological commitments, if they are to reach mature and sustainable expression, need to be grounded in a sense of deep reciprocity with the living world.

Scholars such as Timothy Morton and Bruno Latour (*Nature* 2017) remind us that viewing the natural world as something separate and 'other than' humans is not only ethically problematic but empirically false. For example, microorganisms in our gut aid digestion, while others compose part of our skin. Pollinators such as bees and wasps help produce the food we eat, while photosynthetic organisms such as trees and phytoplankton provide the oxygen that we need in order to live, in turn taking up the carbon dioxide we expel.

5.2 The New Ethic

Taking a moment to look back in time, it was Christmas Eve 1968 and the first crewed mission to the moon had reached its destination. As Apollo 8 entered into lunar orbit the

5.2 The New Ethic

Fig. 5.1 Image of the earth rising over the moon from Apollo 8. (NASA astronauts captured this powerful image 50 years ago. SCIENCE & SOCIETY PICTURE LIBRARY VIA GETTY IMAGES)

crew prepared to read passages of Genesis for a television broadcast to the world. But as the command module came around on its fourth orbit around the Moon, a bright blue and white globe suspended in the black above the haunting grey of the moon was visible through the window. Before that moment over 50 years ago, no one had seen an Earthrise.

The photograph of that instant in time became one of the most influential images in history. A driving force of the growing environmental movement, the photograph, which became known as Earthrise, showed the world as a singular, fragile, oasis (Fig. 5.1).

That singular, fragile oasis is presently under a devastating assault. What, then, will be our new vision of an engineering ethic that addresses directly and forcefully the global environmental crises of our time? How can we shift our perspective from viewing the Earth, not as a dead rock with resources to exploit, but as a living system whose health depends on the health of its organs and tissues—its wetlands, forests, seagrass, mangroves, fish, corals, and more?

Berry (2000) offered three foundational properties needed for grounding this new ethic: differentiation, subjectivity, and community. Interestingly, these are the same properties that are present in a Universe viewed through the lens of quantum mechanics.

When we exam the ongoing evolutionary processes, it is clear that the Universe's intention is to produce a wide variety in all things—from the myriad number of atomic structures to the countless species of plant and animals to the appearance of the human. Differentiation is of primary importance in the appreciation of the entire Earth's process. Estimates of animal species alone range from 3 to 30 million distinct species with approximately 40,000 bird species and 5000 mammal species. The have been at least eight distinct types of human species including *Homo soloensis, Homo floresiensis, Homo denisova, Homo rudolfensis, Homo neanderthalensis, Home erectus, Homo ergaster*, and *Homo sapiens*. Of these eight human species, only one has survived: *Homo sapiens*, us. The cosmologist, Brian Swimme wrote,

> This is the greatest discovery of the scientific enterprise: You take hydrogen gas, and you leave it alone, and it turns into rosebuds, giraffes and humans. *(Hidden Heart of the Cosmos* 2019)

The Earth constantly and continuously evolves from the simple to the more complex in a dynamic, ongoing process characterized by both downward and upward causation.

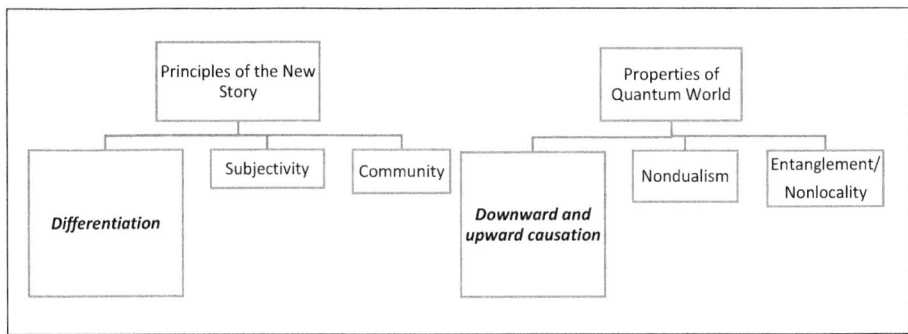

Similarity of New Story and Quantum World Properties: Differentiation and Downward and Upward Causation

According to Thomas Berry, the second important foundational value is subjectivity. Herein, the notion of Nature as "lumps of wax" is rejected and replaced by the recognition that each life form is said to have its own interiority, its own self, its mystery, its numinous aspect. Thomas Berry viewed subjectivity as an essential dimension of the entire evolutionary process and thought that to deprive any being of this quality is to disrupt the total order of the Earth. This principal is consistent with the quantum mechanics property of nondualism.

The philosopher David Loy defined nonduality in the following way: "Two things we have understood as separate from each other are in fact not separate at all. They are so dependent on each other that they are in effect two different sides of the same coin." (*Nonduality: A Study in Comparative Philosophy* 1997) Swimme added, "A tree is a self: it is 'unseen shaping' more than it is leaves or bark, roots or cellulose or fruit. What this means is that we must address trees as we must address all things, confronting them in the awareness that we are in the presence of numinous mystery." (*Hidden Heart of the Cosmos* 2019).

The danger of removing the sense of self or interiority in each living form has resulted in humans beings making, "a terrifying assault on the Earth with an irrationality that is stunning in its enormity, while we were being assured that this was the way to a better, more humane, more reasonable world." (Berry, *Evening Thoughts* 2015).

5.2 The New Ethic

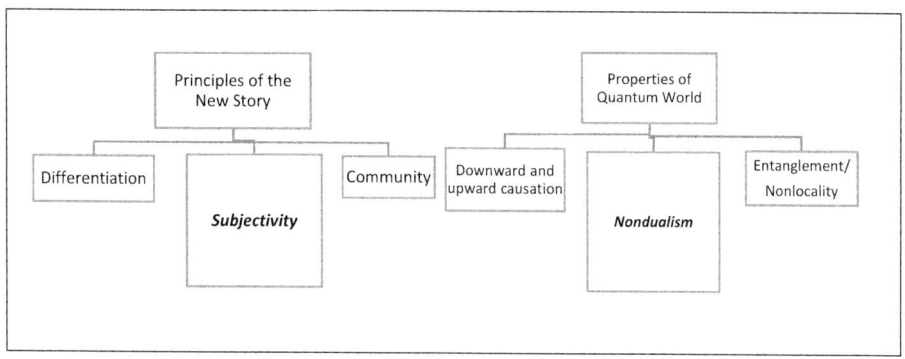

Similarity of New Story and Quantum World Properties: Subjectivity and Nondualism

The third foundational principle of the new story or ethic is communion or community. Community is defined as a group of interdependent organisms of distinct species growing or living together in a specified habitat. The notion of community points to the interconnectedness of all living system on Earth. This property is equivalent to the property of nonlocality in quantum mechanics which describes the apparent ability of objects to instantaneously know about each other's state, even when separated by large distances (potentially even billions of light years). Nonlocality occurs due to the phenomenon of entanglement, whereby particles that interact with each other become permanently correlated, or dependent on each other's states and properties, to the extent that they effectively lose their individuality, and in many ways, behave as a single entity.

It is interesting that, in addition to being an important property in the quantum mechanics world, the concept of nonlocality is found in wisdom traditions throughout history. One tradition that seems particularly relevant is the following attributed to Chief Seattle (2020):

> The Earth does not belong to man; man belongs to the Earth. All things are connected like the blood that unites us all. Man did not weave the web of life; he is merely a strand in it. Whatever he does to the web, he does to himself.

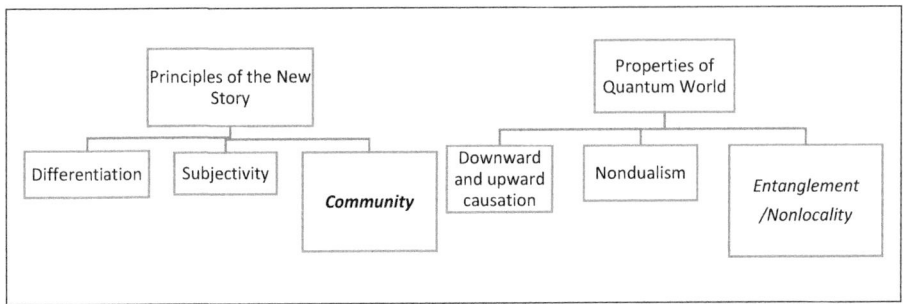

Similarity of New Story and Quantum World Properties: Community and Entanglement/Nonlocality

The combination of the three foundational elements developed by Berry, (differentiation, subjectivity, and community) and the newly gained insights from quantum mechanics establishes the foundations for a new engineering ethic to guide our actions as we address the present environmental crisis.

What then are the fundamental issues that must be included in the new engineering ethic? At a minimum, it must include consideration of the following items:

- A recognition of the existence of both downward and upward causality.
- A preference for differentiation over homogenization, diversity over uniformity
- A recognition of the deep interconnectedness of all life forms and the explicit effects each life has on every other life.
- An awareness of and respect for the intricate web of life that exists throughout the natural world.
- A recognition of the subjective nature of all our observations and experiments and the influences that such subjectivity has on our results.

In his classic essay, "The Land Ethic," published posthumously in *A Sand County Almanac* (1949), Leopold proposed that the next step in the evolution of ethics is the expansion of ethical considerations to include nonhuman members of the biotic community, collectively referred to as "the land." Leopold states the basic principle of his land ethic as: "A thing is right when it tends to preserve the integrity, stability, and beauty of the biotic community. It is wrong when it tends otherwise."

When we combine Leopold's notion of a biotic community with both Berry's ideas about a "New Story" and the principles of quantum mechanics, the following ethic emerges:

> A thing is right when it honors the Earth as a rich tapestry of interconnected species and honors the dynamic, evolving nature of all life forms present, governed by the principles of differentiation, subjectivity, and community. It is wrong otherwise. The hubris of modeling

the Earth as a collection of objects is replaced by a deep and profound gratitude. The impatient arrogance of the desire for control of the natural world is supplanted by extraordinary caution and acceptance in all things.

5.3 A New Engineering Discipline: *Omnium* Engineering

Integrating these new insights and attitudes into the engineering profession will require a significant broadening and reshaping of our ethical responsibility towards the Earth. As noted previously, such change is quite difficult and only occurs during a crisis. It is clear that we are at such a crisis. Just as life continues to evolve, our ethical view needs to continue to evolve as well. Engineering ethics has changed over the course of the last half century to include limited concerns over the finiteness of natural resources and sustainability, the world's poor and the issues of peace and social justice. I believe that the new ethic stated above which combines the concept of Leopold's biotic community with the implications of quantum mechanics and Berry's "New Story" will result in the creation of a new engineering discipline with an even more expansive wave of ethical responsibility: one that I have described and named *omnium* engineering (Catalano, December 2020).

A literal translation of the Latin word *omnium* is "all" or "all beings" yet in the context of engineering it can be extended to more fully describe an engineering profession that considers the health and well-being of the entire Earth and all its life forms—not only that of the human species (Fig. 5.2).

This new engineering discipline of *omnium* engineering seeks to push the ethical boundaries of the practice of modern engineering to include "all my relations," an interdependent, interconnected community of all forms of life which includes other people, animals, birds, insects, trees, and plants, and even rocks, rivers, mountains, and valleys to borrow from the Lakota people. (Lakota Prayer)

> We are in the circle of life together, co-existing, co-dependent, co-creating our destiny. One, not more important than the other. One nation evolving from the other and yet each dependent upon the one above and the one below. All of us a part of the Great Mystery. Mitakuye Oyasin – The Prayer

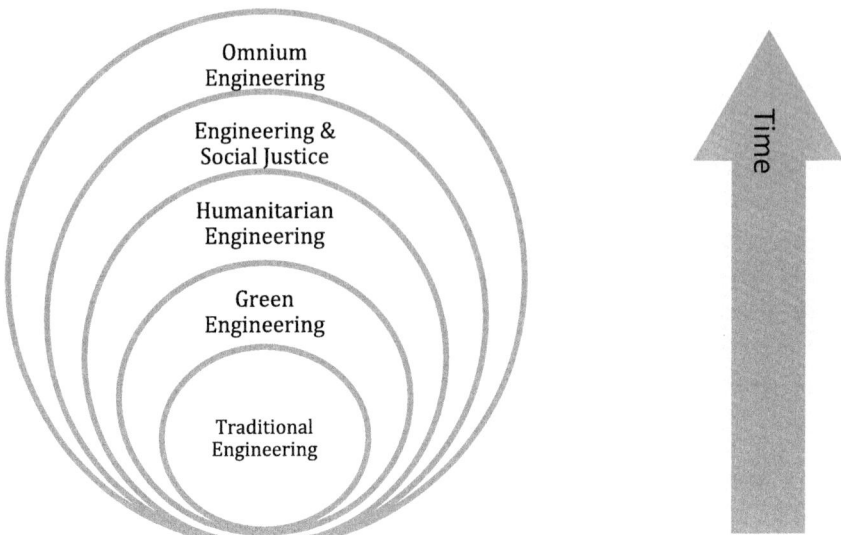

Fig. 5.2 Evolving engineering ethic (Catalano 2022)

6. The Ethical Implications of the Mechanical and Quantum Perspective: A Case Study

> *We do not see Nature with our eyes, but with our understandings and our hearts.*
>
> *William Hazlitt.* (William Hazlitt was an English essayist, drama and literary critic, painter, social commentator, and philosopher. He is now considered one of the greatest critics and essayists in the history of the English language, placed in the company of Samuel Johnson and George Orwell)

6.1 History of Wolves in Yellowstone National Park

One of the important consequences of global climate change is the rapidly increasing rate of plant and animal species extinctions. As noted at the beginning of this text, many scholars are now labelling the present time as the Sixth Extinction. While there are a wide variety of different components involved in the extraordinarily complex issue of extinction, one important aspect that we will focus on in this chapter is the modeling required in the management and control of endangered species, typically within a specific geographic location. Because this modeling has historically utilized a classical mechanics approach along with the lens of scientific materialism, we can use it to reflect upon the impact of the proposed new engineering ethic and how that ethic based on the quantum world view would change the way engineering approaches environmental issues. In this chapter we will consider the management and control of one endangered species, the wolf, in one region, Yellowstone National Park, not only from a classical mechanics perspective but also from a quantum perspective. This comparison provides a richer understanding of the ethical foundation underpinning the two different scientific paradigms.

In the 1800s, westward expansion in the United States brought settlers and their livestock into direct contact with both native predator and prey species. One of the most hated species during this time was the wolf. We frequently hear two explanations for why wolves are so feared, one being the folklore and fairy tales ("Little Red Riding Hood," "The Three Little Pigs," etc.) that we inherited from Europe; and the other being the fact that wolves kill livestock and compete with humans for wild game. What is clear is that wolves became a focal point for management, control and ultimately elimination.

Predator control which included poisoning was practiced throughout the western U.S. in the 1800s and early 1900s. While other predators such as bears, cougars, and coyotes were also killed to protect both livestock and "more desirable" wildlife species such as deer and elk, no species was killed with the intensity and ferocity focused upon wolves (Fig. 6.1).

The gray wolf was present in Yellowstone when the park was established in 1872. The *Yellowstone National Park Act of 1872* stated that the Secretary of the Interior "shall provide against the wanton destruction of the fish and game found within said Park." Unfortunately, the park management considered the wolves' habit of killing prey species "wanton destruction" of park animals and determined set about to eliminate them. It is important to note that this was an era before the concepts of 'ecosystem' and the 'interconnectedness of species' were recognized or understood. Between 1914 and 1926, at least 136 wolves were killed in the park; by the 1940s, wolf packs were rarely reported. By the mid-1900s, wolves had been almost entirely eliminated not only in Yellowstone Park but also throughout the lower 48 states in the U.S.

Removing wolves from the park had an enormous impact on the park because we now understand that wolves are a 'keystone' species, i.e., a species which has a disproportionately large effect on its natural environment relative to its abundance. This concept, introduced in 1969 by the zoologist Robert T. Paine. (*Food Web Complexity and Species Diversity* 1966), plays a critical role in maintaining the structure of an ecological community, affecting many other organisms in an ecosystem, and helping to determine the types and numbers of various other species in the community. Without a keystone species, an ecosystem becomes dramatically different or in some ceases to exist altogether.

We now know that predators are especially important to an ecosystem because they control population numbers of other species. In Yellowstone before their elimination wolves typically fed on elk which controlled the number of elk in the park; after their elimination, the elk population numbers exploded. Because there were more elk, the elk began to feed on younger aspen trees, which resulted in the number of aspens declining. Without the predation of wolves, elk behavior changed, with elk herds often remaining in one place and feeding on vegetation near the rivers. With significantly less vegetation, the riverbanks eroded, and the rivers widened. In the warmer summer months, the temperature of the rivers rose significantly because there was no shade to cool the water, which in turn affected the abundance and distribution of fish species. Birds that nested by the river were also affected because riverbank was devoid of the vegetation that the birds

6.1 History of Wolves in Yellowstone National Park

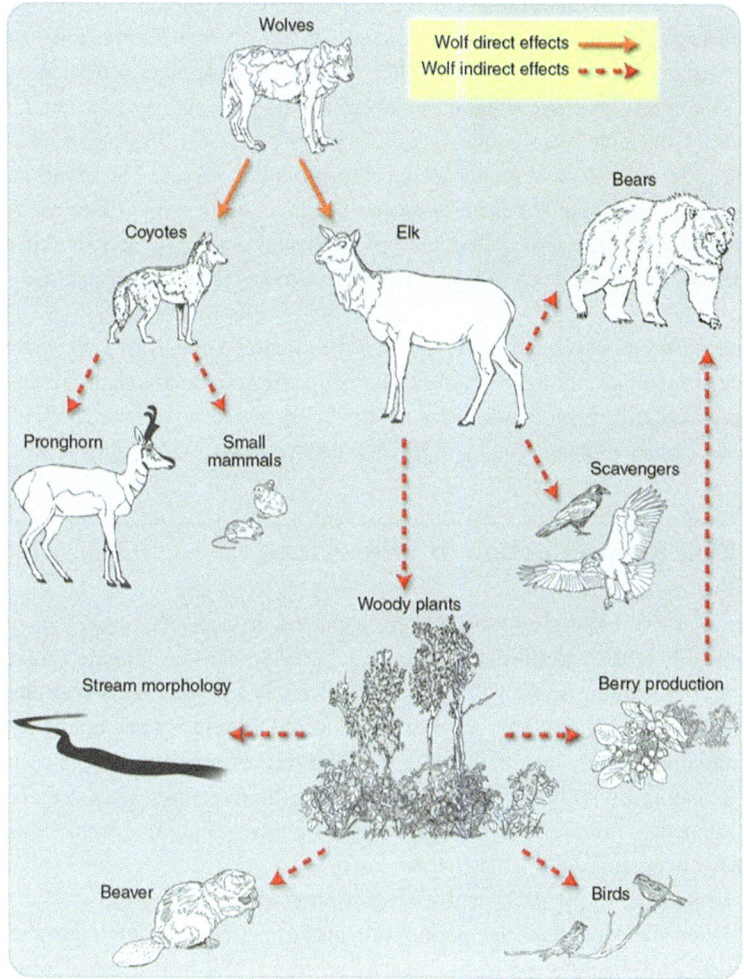

Fig. 6.1 Conceptual diagram showing effects of gray wolf reintroduction into the Greater Yellowstone Ecosystem (Ripple 2014)

depended upon which to build their nests. Before the elimination of wolves, beavers had used willow trees for their dams, but after the elimination of wolves, there were fewer willow trees along the river because of the overabundance of elk; as a result, eventually the beavers disappeared. The elimination of wolves also resulted in higher coyote population densities because the wolves' absence opened a niche that coyotes could partially occupy in Yellowstone. The increase in coyote population, in turn, greatly reduced the rodent population which impacted food availability for a wide range of smaller predators.

Attitudes towards the natural world began to change beginning in the 1960s. With the publication and widespread success of Rachel Carson's *Silent Spring*, a national awareness of environmental issues and their potential consequences led to the passage of laws designed to correct the environmental mistakes of the past and prevent similar mistakes in the future. One such law was the *Endangered Species Act (ESA)*, passed in 1973. This act required the federal government to develop programs for the conservation of threatened and endangered plants and animals and the habitats in which they are found. The *Endangered Species Act* also called for governmental action to correct the population management mistakes made in the past and restore ecosystems to their natural state when possible.

For almost 70 years scientists had observed the Greater Yellowstone ecosystem without the presence of wolves, noting the significant changes described above. Realizing that wolf populations were an irreplaceable piece of the ecosystems, government scientists began the process of reintroducing wolves to the Greater Yellowstone Ecosystem in 1995.

6.2 Wolf Reintroduction in Yellowstone

In January of 1995, eight gray wolves were captured in Jasper National Park in Alberta, Canada, and brought to Yellowstone National Park. Scientists continued to reintroduce gray wolves by bringing small packs to Yellowstone from other areas with healthy wolf populations, with a total of thirty-one wolves relocated to Yellowstone by the end of 1996. Since the reintroduction, scientists and a wide variety of scholars have studied a range of topics including individual wolf behavior, population dynamics, wolf-prey relationship dynamics, genetics, disease, and management and public policy questions. Many of their findings have been completely unexpected.

Upon their return to the park, wolves began preying upon the elk that were damaging the willows and aspens near the park's streams. The trees began to grow again, and with the increasing healthy forest, the beavers returned. The beaver population increased which resulted in the building of new dams and ponds which have multiple effects on stream hydrology including moderating the seasonal pulses of runoff, storing water for replenishing the water table, and providing cold, shaded water for fish. With less erosion along the riverbanks, the rivers also began to maintain their course. In addition, new robust willow stands appeared and provided habitat for countless songbirds. The presence of wolves in the park also reduced stress levels among deer which helped prevent disease from spreading within the herds.

In the last decade, scientists noted that winters in Yellowstone had been less severe. In the absence of hard winters, wolves tend to be the primary reason for elk mortality. Today researchers have determined that the combination of less snow and more wolves has provided many benefits for a wide array of scavengers, including ravens, eagles, magpies, coyotes, and grizzly and black bears emerging from hibernation. Instead of

6.2 Wolf Reintroduction in Yellowstone

a highly fluctuating cycle of elk carrion availability, as existed before wolves and when winters were harder, there is now a more reliable distribution of carrion throughout winter and early spring.

A summary list of changes in Yellowstone since wolf reintroduction in 1995 is shown in Table 6.1.

Today there are only 95 wolves left in the park compared to the estimated 528 wolves who resided in the Greater Yellowstone Ecosystem as of 2015. Over the course of 2021 alone, the wolf population decreased 23%. In addition, federal wildlife officials removed the gray wolf from the U.S. Endangered Species Act list in October 2020, claiming the wolf population has recovered and the animal no longer needs federal protection. With removal from the ESA list, state governments now have the authority to manage gray wolf populations and prevent livestock losses in accordance with state law. Over 3500

Table 6.1 A summary of changes in Yellowstone after wolf reintroduction in 1995 (Peterson 2020)

Deer and other ungulates	While wolves killed deer, diminishing their population, wolves also changed the deer's behavior. When threatened by wolves, deer did not graze for extended periods in the same area. Because they moved around more, the herd aerated the soil, allowing the grasslands an opportunity to recover
Grass and trees	As a result of the deer's changed eating habits, the grassy valleys and forests regenerated more quickly. Trees in the park grew to as much as five times their previous height in only six years
Birds and bears	Newer and larger trees provided a place for songbirds to live and produced berries for bears to eat. The healthier bear population then killed more elk, contributing to the cycle the wolves started
Beavers and other small animals	Larger trees and healthier vegetation also allowed beaver populations to flourish. Their dam building habits provided habitats for muskrats, amphibians, ducks, fish, reptiles, and otters
Other carnivorous mammals	Wolves also killed coyotes, thereby increasing the populations of rabbits and mice. This created a larger food source for hawks, weasels, foxes, and badgers
Scavengers	Ravens and bald eagles fed off of the larger mammals' kills
The land	Soil erosion of riverbanks caused greater variation in the path of the rivers and creeks. But, as a result of elk herds being less stationary and more vegetation growing next to rivers, riverbanks stabilized. Thus, wolves not only changed Yellowstone's physical geography but also the course of the rivers and creeks

wolves have been killed by state government agents and private citizens alike in the past few years alone.

Environmental groups called the delisting a tragic backward step in the face of global crisis, with a changing climate, declining biodiversity and a rising number of plant and animal extinctions. They believe delisting is a death sentence for the gray wolf, whose recovery is still in its infancy.

6.3 Wolf Management: A Comparison of Old and New Scientific and Ethical Paradigms

Wildlife management is the science of reaching goals by manipulating and/or maintaining wildlife habitats and populations. This process involves many components including knowledge and understanding of wildlife population trends; factors that influence wildlife populations; interaction of wildlife species; impact of humans; and how surrounding landscape affects wildlife. It states that, as a profession, it attempts to balance the needs of wildlife with the needs of people using the best available science.

I disagree. I believe that, in fact, wildlife management does not use the best available science. Rather it is based on a scientific model and ethical view of the Universe that is outdated—the 'Nature as machine' paradigm. Ultimately, the goal of traditional wildlife management is to determine numbers, i.e., the carrying capacity which is the maximum population size of a biological species that can be sustained by that specific environment, given the food, habitat, water, and other resources available. The effect of carrying capacity on population dynamics is typically modelled with a logistic function. The logistic function, a special kind of exponential function which typically models the growth of a population, considers the 'carrying capacity' of land. Even the concept of 'carrying capacity' is a relic from the past as the term evokes images of trucks, trains and airplanes loaded to the maximum allowable limit and transporting 'goods' to market. It is a concept linked directly to the mechanical world view and scientific materialism.

In the mechanical Universe, there are typically three different approaches to modelling the interactions between populations and available resources. The simplest approach is the Lotka Volterra or predator prey model. The Lotka–Volterra equations are a pair of first-order nonlinear differential equations, frequently used to describe the dynamics of biological systems in which two species interact, one as a predator and the other as prey. Note the interactions are limited to typically two species which is hardly realistic. Ecosystems are filled with countless numbers of species, each interacting in countless ways.

The second approach is the Monte Carlo method. This method, a broad class of computational algorithms, relies on repeated random sampling to obtain numerical results. The underlying concept is to use randomness to solve problems that are deterministic in principle. Note the requirement that the problem must be deterministic. Ecosystems are filled with random, chaotic occurrences from freak storms to infestations of non-native bacteria.

The third approach is the simulation method. A simulation is the imitation of a real-world process or system over time. Simulations require the use of models; the model represents the key characteristics or behaviors of the selected system or process, whereas the simulation represents the evolution of the model over time. Note that this approach requires the use of models which are most often deterministic and limited to a relatively small number of processes. These restrictions are extremely limiting at best and probably lead to incorrect results.

None of these models have proven successful in providing guidance in wolf management primarily because such models of Nature view not only wolves but all of Nature as a "machine"—a collection of objects. In the mechanical Universe view, the wolf acts and reacts in the same way a mechanical spring in a clock behaves when compressed. The wolf then is perceived as if it is a separate part of a lifeless machine that can somehow be manipulated and ultimately "managed" according to the whims/preferences of the times. It is afforded the same ethical consideration as a part in that machine or in the words of Descartes, "a lump of wax."

Using the mechanical world model to consider this particular lump of wax, the wolf, then we must believe that:

- Wolves individually are indistinguishable one from another; their traits are identical and are unchanging.
- Wolf pack structure and its allocation of distinct roles for the different wolf pack members does not exist
- Wolves are simply killing, defecating reproducing robots who interact with an extremely limited number of prey species.
- Wolves react from instinct only; they do not possess sentience, feel happiness or sadness, love, or loss.
- Wolves have no impact upon non-prey species nor the local environment in which they live.

The wolves modelled in the mechanical Universe simply do not exist. Moreover, the natural world that is modelled using the classical world view does not exist. I am reminded of an infamous problem often cited in classical mechanics. Dubbed the "spherical cow" problem, this problem is a humorous metaphor for highly simplified scientific models of complex phenomena. Originating in theoretical physics, the metaphor refers to tendency of both physicists and engineers to reduce a problem to the simplest form imaginable in order to make calculations more feasible—even if the simplification hinders the model's application to reality (Shelton and Cliffe 1999). No, physicists and engineers do not think that cows are spherical; we know this is a ridiculous approximation. Yet the approach remains firmly embedded in our view of the natural world (Fig. 6.2).

Fig. 6.2 A pasture filled with spherical cows (Shelton and Cliffe 1999)

6.4 Wolves and the Quantum Universe

How might wolves be viewed in the paradigm described by quantum principles? How would we characterize them in light of the new insights provided by quantum mechanics and the ethical principles described by Berry?

- Entanglement/connectedness

Wolves play an integral role throughout the entire Yellowstone ecosystem. The wolf packs provide a stable source of carrion not only for themselves but also for a range of different carnivorous animals. Wolves dramatically reduce the coyote population which, in turn, ensures the availability of rodents as food sources for other predators. Apart from their own pack, wolves interact with ravens more than any other animal. It appears that ravens are the most obvious beneficiary from the wolf-raven relationship. Studies have found that 100% of wolf kills are visited by ravens, and the raven consumes approximately two thirds of the carcasses.

- Nonduality

A wolf pack exhibits behaviors and relationships as rich as a human family.

Wolves feel great attachment to their pack members, with all members caring for the pups, and honoring and feeding the elders of the pack. Wolves are known to have emotional lives, experiencing emotions such as joy and grief.[1]

---

[1] After the death of a wolf, the remaining members of the pack typically walk with their heads and tails held low—a sign of depression. They no longer howl as a group, but each one cries in their own way. This behavior often lasts for a few weeks. Jim and Jamie Dutcher describe the grief and mourning in a wolf pack after the loss of the low-ranking omega female wolf, Motaki, to a mountain lion

6.4 Wolves and the Quantum Universe

- Self-organization and emergence

Living in family groups called "packs," wolves are a well-organized species with a social hierarchy that defines the role and contribution of each member. There is typically an alpha pair. The leaders not only make decisions but also take actions to benefit the whole pack while allowing others to perform duties for the pack that align with their strengths. There is demonstrated respect for their elders as the rest of the pack learns from their experience. Additionally, the sick or weak wolves are cared for by other member of the pack. Roles within the pack are sometimes shared, giving others the opportunity to help the entire group be stronger, healthier, and more resilient. Roles also change in time with the emergence of new leaders.[2]

- Downward causation

After 70 years without wolves, the reintroduction caused unanticipated change in Yellowstone's ecosystem and even its physical geography. One type of process referred to as trophic cascade in ecology occurs when the food web is disrupted. Trophic cascade, is an ecological phenomenon triggered by the addition or removal of top predators and involving reciprocal changes in the relative populations of predator and prey throughout a food chain, often resulting in dramatic changes in ecosystem structure and nutrient cycling. For example, this kind of a top-down cascade occurs when wolves are effective enough in predation to reduce the abundance, or alter the behavior of their prey, thereby releasing the next lower trophic level from predation (or herbivory if the intermediate trophic level is an herbivore). Figure 6.3 depicts the key features of a trophic cascade and, in addition, offers a glimpse into the other effects that result from the elimination of wolves—fewer large predators, fewer species of plants, diminished forest cover, less diverse songbirds and numerous large herbivores.

(Dutcher 2019). The pack not only lost their spirit and their playfulness; they also no longer howled as a group, but rather "sang alone in a slow mournful cry." They were depressed—tails and heads held low and walking softly and slowly—when they came upon the place where Motaki was killed. They inspected the area and pinned their ears back and dropped their tails, a gesture that usually means submission.

[2] In Rick McIntyre's recent trilogy, the wolf numbered 302 is a lover, not a fighter. In his early years, he was a renegade with an eye for female wolves, often abandoning females he had gotten pregnant. He also often fled from danger, and even napped during a heated battle with a rival pack. He also begged for food from other wolves As McIntyre closely observed, wolf 302 began to mature, and much to McIntyre's surprise, became the leader of a new pack in his old age. In a year when game was scarce, the aging wolf provided for his family and changed enough to live up to the legacies of the alpha males before him.

Fig. 6.3 Trophic cascade (Encyclopædia Britannica)

- Upward causation

Recalling that upward causation refers to smaller systems clustering to form larger systems or conglomerates, it is clear that Yellowstone is filled with examples of upward causation linked to wolves. One such example can be seen in the structure of wolf packs as they begin a hunt. The pack is led by one of the alpha wolves with the remainder of the wolves in line behind and the other alpha wolf at the rear. *The pack then moves according to the elders' pace with all wolves on high alert. Once the prey is identified, the old, sick, and young wolves stay behind as the alpha pair and younger wolves begin the attack.*

6.5 Approaching a Decision

Wolves are complex highly intelligent animals who are caring, playful, and above all devoted to family. Only a select few other species exhibit these traits so clearly. Wolves educate their young, take care of their injured, and live in strong family groups or packs. A wolf pack is an exceedingly complex social unit; it is an extended family of parents, offspring, siblings, aunts, uncles, and sometimes dispersers from other packs. There are old wolves that need to be cared for, pups that need to be educated, and young adults that are beginning to assert themselves—all altering the dynamics of the pack.

When we reflect on wolves from the quantum world perspective, we see that they are an integral thread in the rich tapestry of the natural world, the same world in which we also belong. Their presence impacts not only members of prey species but also the course

6.5 Approaching a Decision

of rivers and streams. We are forced to wonder how many more effects have we yet to identify? And what are their effects on us, as humans, on our culture and our story?

We began this section of the text exploring the ultimate wildlife management question: how many individuals are the 'correct' number for the carrying capacity of a portion of the land? More specifically, we have been concerned with the question: how many wolves should be allowed to live in Yellowstone National Park before they become managed, controlled, and ultimately deemed expendable?

We have seen that in the mechanical Universe paradigm using the lens of scientific materialism, scientists select a set of mathematical tools, complete the calculations, and arrive at a fixed number with little if any thought given to the myriad number of possible consequences involved and certainly even less ethical concern for any lives lost.

The questions of the appropriate "wolf numbers" and the reasonable "carrying capacity" of the land will never be answered with certainty by ever more elaborate mathematical equations and computational algorithms that are based on a model of the world whose time has passed. If we admit that such certainty does not exist, then how can we approach such questions in the future?

I believe that a quantum mechanics world view offers a path. If we consider the question of "carrying capacity" as well as other questions regarding the natural world in light of quantum mechanics principles as a guide, we begin to recognize the intricate and entangled interconnectedness of Nature; it is clearly apparent that both downward and upward causation are ever present along with the emergence of an infinite number of unforeseen possibilities.

Using the quantum viewpoint, we as engineers, begin to see ourselves as much more than clever watchmakers tinkering at the margins of life itself, and, as a result of that shift in perspective, we see the wolf as much more than an eating, defecating, reproducing robot devoid of all sentience.

Our profession has impacted the Earth in profound ways since the Industrial Revolution and the subsequent adoption and integration throughout our culture of a mechanical world view. Earlier I quoted from Aldo Leopold's *A Sand County Almanac*. In his essay, "Think Like a Mountain." Leopold suggested that to think like a mountain is to have a complete appreciation for the profound interconnectedness of all the elements in the ecosystems. Such a view, which is consistent with modern science, demands from us a radical[3] sense of awe and humility as we confront the unfathomable mysteries of the natural world.

---

[3] The adjective 'radical' is chosen purposefully as evidenced by its various meanings: arising from or going to a root or source; departing markedly from the usual or customary; extreme or drastic; relating to or advocating fundamental or revolutionary changes in current practices, conditions, or institutions; and lastly, grounded in the Earth.

7. Final Reflections

7.1 Intergovernmental Panel on Climate Change

I began this text quoting from an article published in 2004 in *National Geographic* that warned that the Earth was in crisis. In early 2022, an extensive review was published by the Intergovernmental Panel on Climate Change (IPCC) entitled "The Climate Change 2022 Impacts Report." This panel is composed of the world's leading climate scientists who are charged with publishing regular comprehensive updates of global knowledge on the climate crisis with the purpose of informing government policymaking. The 2022 report includes warnings that make for grim reading. They include:

- 3.5 billion people are highly vulnerable to climate impacts.
- Half the world's population currently suffers severe water shortages at some point during each year.
- Increased number of people are at risk of serious flooding each year.
- One billion people living on coasts will be at risk of being displaced by 2050.
- One in three people are exposed to deadly heat stress each year—a percentage that is projected to increase to 50–75% by the end of the century.
- Rising temperatures and rainfall are increasing the spread of diseases not only in people but also in crops, livestock, and wildlife.
- Even if the global average temperature increase is kept below 1.6 °C by 2100—and we are already at 1.1 °C—8% of today's farmland will become climatically unsuitable. This will occur at the same time the Erath's population is forecast to exceed nine billion.
- If global heating continues and few changes are made, more than 183 million people are projected to be without adequate nutrition by 2050.

- Animals and plants are being exposed to climatic conditions not experienced for tens of thousands of years. Half of the studied species have already been forced to relocate, and many face extinction.
- Maintaining the resilience of Nature at a global scale depends on the conservation of 30–50% of Earth's land, freshwater, and oceans. Today, less than 15% of land, 21% of freshwater and 8% of oceans are protected areas.
- Some regions, like the Amazon, have switched from storing carbon to emitting carbon.

7.2 Other Issues

Although the focus of the present text has centered on the case of Yellowstone wolves and the issues associated with managing the natural world, this is just the leading edge of a wide range of ethical issues that span the breadth of engineering. Here we will limit our discussion to two newer ethical dilemmas which engineering faces today. The first ethical issue concerns the actual nature of life on Earth and developments in manipulating life's genetic make-up. (Synthetic biology) The second ethical issue is linked to the Earth itself, specifically the proposed modification of the Earth's atmosphere to combat rising temperatures (Geoengineering).

Synthetic biology
Synthetic biology, as noted earlier in this text, is a multidisciplinary area of research that seeks to create new biological parts, devices, and systems, or to redesign systems that are already found in Nature. It is a branch of science that encompasses a broad range of methodologies from various disciplines including biotechnology, genetic engineering, molecular biology, molecular engineering, systems biology, membrane science, biophysics, chemical and biological engineering, electrical and computer engineering, control engineering and evolutionary biology. Rapid advances in synthetic biology such as the development of genome editing are forcing engineers to confront the difference between a mechanical view of life versus a quantum view head on.

Genome editing technologies have led to fundamental changes in genetic science. Among them, CRISPR-Cas9 technology particularly stands out due to its many advantages, including easy handling, high accuracy, and low cost. It has been introduced in fields related to humans, animals, and the environment, while presenting puzzling questions, applications, concerns, and bioethical issues to be discussed. One of the important resultant ethical concerns focuses upon ecological imbalance that may result.

In studies using RNA-targeted gene editing methods based on CRISPR-Cas9, unintended effects are a critical concern. Since gene drift will persist in a population, possible unintended or off-target mutations will continue in each generation. In addition, the number and effect of mutations may increase as generations progress (Rodriguez 2016).

Another concern is the possibility that genes can be transferred to other species in the environment which has the potential to result in the transmission of negative characteristics to the associated organisms (Esvelt 2019). The distribution of the properties of the entrained genes among the populations can make control exceedingly difficult.

Each of these issues in turn requires engineering to make a choice concerning the Universe and the Earth. Do we live in a deterministic, lifeless Universe to be molded according to our preferences? Or do we live in a web of life filled with connections, and possibilities? I submit that the new scientific paradigm of quantum mechanics demands the latter response.

Geoengineering

A National Academies of Science, Engineering and Medicine report released in March recommends that the United States invest unilaterally to expand research on solar geoengineering, a set of controversial proposed strategies to cool the planet by reflecting sunlight back to space (National Academies of Sciences 2021). However, unilateral, preemptive research without broad public participation, and before a global governance structure is established, risks exacerbating international conflict and undermining progress on energy system transformation away from fossil fuels in the highly contested and politicized landscape of global climate policy (Stephens 2020).

The idea of a technical intervention to counter global warming may have some appeal, but the social, political, and environmental risks associated with solar geoengineering research are yet to be considered (Trisos et al. 2018). Given the dangers of advancing solar geoengineering (Abatayo et al. 2020) including further concentrating power among the more affluent parts of society (Stephens 2020) and deterring mitigation efforts (McLaren 2021) inclusive processes for public deliberations on whether, when, and how public funding should be provided to support climate manipulation are essential. More diverse voices are needed to expand public discussion beyond the often-narrow narrative that limits authentic deliberation about the risks.

Two ideas presently proposed in geoengineering are carbon dioxide removal (CDR) and solar radiation management (SRM).

Carbon dioxide removal (CDR) faces limitations to its effectiveness. The IPCC has found that CO_2 removals are partially counteracted by CO_2 release from oceans and land reservoirs. Uncertainties around permanent storage further undercuts CDR's effectiveness in reaching temperature goals. There are tremendous risks involved with CDR including the possibility of triggering tipping points and feedbacks such as permafrost thawing and forest ecosystem degradation. These possibilities would drastically increase at warming levels beyond 1.5 °C.

Finally, CDR fails to address many of the other dramatic changes in the global climate system such as rising sea levels that threaten the existence of small island states and low-lying coastal areas, and the millions of people in those areas. Situations like these would continue for centuries to millennia.

Solar radiation management (SRM) also comes with severe risks such as the disruption of regional and seasonal rainfall patterns, ozone depletion, ocean acidification, and other known and unknown adverse side-effects. Because SRM like CDR would only "mask" temperature rise and not address the root cause of the problem, a sudden termination of the efforts would induce a rapid acceleration of climate change—a so-called "termination shock." The Earth's climate is changing faster and in ways it never has before, so the question really becomes: How do you maximize ecosystems' ability to cope with change? This is especially critical if change occurs abruptly. Adaptation would be impossible for countless species and ecosystems.

Each of these issues as well as many others present in geoengineering requires the engineering profession to decide again concerning our relationship to the Universe and the Earth. In each case, we are forced to answer the following questions: Are we master designers who think we can replace parts and twist knobs to adjust the Earth's atmosphere to our liking? Or do we consider the vast potential for both upward and downward causation and the possible unintended consequences that can result from modifying the nature of the Earth's fundamental processes? The insights revealed in quantum mechanics emphatically require the second response.

7.3 Moving Forward

Humans have, for many thousands of years, had measurable impact on the Earth through fire, agriculture, and urbanization. Since the Industrial Revolution the scope, scale, and speed of such impacts have not only exceeded all those in the past, but they are on course to become even more dramatic in the future. Because it is engineering and engineers that are primarily responsible for these impacts, it is we as engineers who must act in response to the Earth in crisis.

I have argued that our engineering ethical codes are prisoners of a set of assumptions about the Earth, and by extension the Universe, which are no longer consistent with the model that modern advances in science have provided. No one can deny that today we live in a quantum world where quantum principles not only explain how matter behaves at the subatomic level but are also used to create the technology and devices that we depend upon for almost every aspect of our everyday lives. Lasers, transistors, GPS, and mobile phones are everywhere with the next wave of innovation promising both unbreakable encryption and computers that are up to one million times faster than today.

Because the quantum model suggests that infinite possibilities for both good and evil exist, this perspective demands that we, as engineers, use extraordinary caution in all decisions. In the quantum world, cause and effect no longer can be used as a "tool" for judging the desirability of a particular ethical choice because, in fact, ethical decision-making is never that simple. Determinism and certainty are rarely if ever relevant or even possible. Stability and returning to the equilibrium state of homeostasis are illusory as we have discovered in our well-intentioned efforts to return the natural world to stability and equilibrium after oil spills and nuclear accidents. Quantum principles inform us that change occurs in both downward and upward directions and that ever-present change is the normal, ongoing condition.

7.4 The Choice Ahead

Growing up in the 1950s, on our nascent television programs, Native American peoples were depicted as speaking using only one-word guttural noises. We could never have imagined that they actually possessed deep wisdom and insight into the natural world. They spoke of the interconnectedness of all things long before the advances of quantum mechanics. They cautioned against short-sightedness and narrow focus in making decisions that affect the Earth. Rather they spoke of the Seventh Generation.

The "7th generation" principle taught by Native Americans says that in every decision, be it personal, governmental, or corporate, we must consider how it will affect our descendants seven generations into the future so that the pristine sky, field, and mountains will still be here for them to enjoy. Long before scientists challenged us to consider "carbon footprints" and "sustainability," Indigenous peoples lived in balance with the world around them. Native American tribes did not even have a word for "ecology"—respect for the Earth was so fundamental in their lifestyle that one word would be too limiting.

I have stated many times throughout this text that we, as engineers, are confronted with a choice. Our choice is similar to the choice that the Lakota people speak of—one that each person must make at some point in life. The Lakota speak of the Red and Black Roads (Marshall 2001). The Black Road is considered the easier road because it avoids obstacles and thus avoids personal growth. The more difficult path is the Red Road in which the individual is confronted with challenges and dangers which, although difficult, have the potential to result in personal transformation.

I believe the easier path in engineering, the Black Path, is to continue blindly adhering to the values and views based on a "mechanical world" view. Even though we have expanded our ethical sphere of responsibility ever so slightly, we have failed to question the scientific model of Nature seen through the lens of scientific materialism, a model in which the natural world has no intrinsic value, and which underlies so many of our professional decisions. Our conditioning through our education has made this our foundation for making professional decisions.

The more difficult path for us to traverse is to consider a new and more inclusive world view that results from the advances in quantum mechanics. In this text I have offered a new ethic:

> A thing is right when it honors the Earth as a rich tapestry of interconnected species and honors the dynamic, evolving nature of all life forms present, governed by the principles of

differentiation, subjectivity, and community. It is wrong otherwise. The hubris of modeling the Earth as a collection of objects is replaced by a deep and profound gratitude. The impatient arrogance of the desire for control of the natural world is supplanted by extraordinary caution and acceptance in all things.

With this new ethic as the foundation, I am suggesting a new engineering discipline, *omnium* engineering, which rejects the false narrative that the Earth is a collection of lifeless objects and replaces that outdated account with a 'New Story', one that recognizes our planet as an immensely diverse community of resplendent subjects and engineering as a profession that seeks to reduce poverty and injustice in the world, recognizes the intrinsic worth of the natural world and treats all species with respect and compassion (Fig. 7.1).

It is my hope that both will serve our profession as springboards for future discussions focusing upon a rethinking and reformulation of our ethical responsibilities. While breaking with the past and opening ourselves to a new way of viewing the natural world is the more difficult road, it is the road I believe offers hope as we confront the Earth in crisis.

Fig. 7.1 Elements of omnium engineering (Catalano 2014)

Appendix

The principles of *omnium* engineering can be illustrated through the plot of a Sierpinski triangle fractal (Peitgen et al. 1991). A fractal is a curve or geometric figure, each part of which has the same statistical character as a whole. Fractals are useful in not only modeling structures such as eroded coastlines or snowflakes, but also in describing partly random or chaotic phenomena (Fig. A.1).

The Sierpiński triangle is a fractal based on a triangle with four equal triangles inscribed in it. The central triangle is removed and each of the other three treated as the original was, and so forth, creating an infinite regression in a finite space. We can locate one of the foundational ethical principles—differentiation, subjectivity, and connectedness—at the three different vertices of each particular triangle. The physical size of each of the triangles vary from the micro to the macro level.

For example, with the largest scale representing the Earth (Triangle #0), we move down to successively smaller scales, moving from the Earth to individual ecosystems (Triangles #1-12) to individual animals and plants (Triangles #13-39) to the microbial level (Triangles #40-120). Organized in this manner, the fractal triangle clearly illustrates that all life shares these fundamental principles. The Sierpinski triangle could be extended to larger and smaller sizes as well with each still governed by the three foundational principles identified by Berry and reinforced by quantum mechanics.

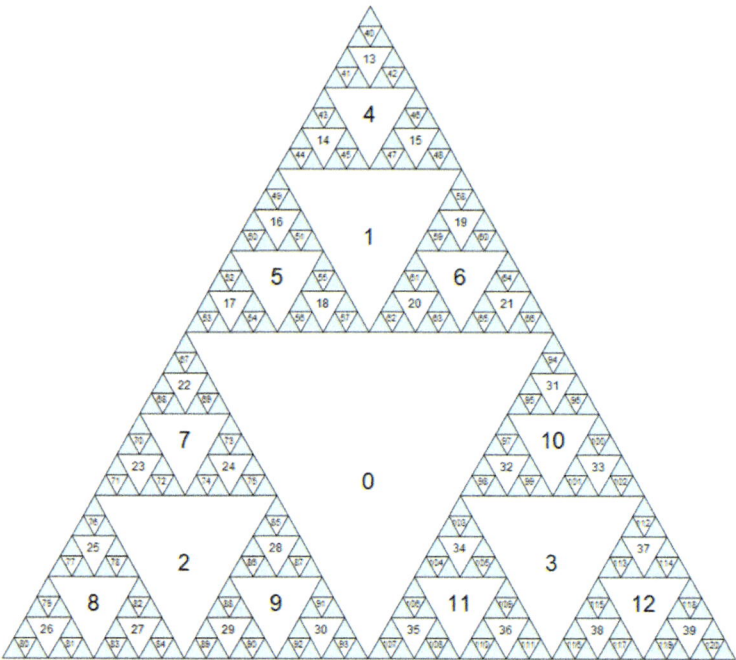

Fig. A.1 Fractal aspect of omnium engineering (Peitgen et al. 1991)

Bibliography

Einstein, B. P., & Rosen, N. (1935). "Can Quantum–Mechanical Description of Physical Reality be Considered Complete?". *Physical Review. 47 (10)*, 777–780.
Anastas, P. (2013). *Innovations in Green Chemistry and Green Engineering.* New York: Springer.
ASME Research Study. (2012). *Learn More about the future of the engineering profession.* ASME.
Augustine, S. (December 2017). *The City of God (Translated with an Introduction by Marcus Dods).* New York: Digireads.
Bacon, F. (July 2008). *The Major Works.* London: Oxford University Press.
Baillie, C. (2012). *Heterotopia: Alternative pathways to Social Justice.* John Hunt Publishing.
Berry, T. (2000). *The Great Work.* New York: Crown Press.
Berry, T. (2015). *Dream of the Earth.* San Francisco: Counterpoint Press.
Berry, T. (2015). *Evening Thoughts.* San Francisco: Sierra Club.
Billing, S. (2017). *Scientific Materialsim and Ultimate Conceptions.* New York: Andesite Press.
Buescher, C. (2018). *History of Environmental Engineering. Retrieved from Washington University in Saint Louis.* https://eece.wustl.edu/eeceatwashu/about/Pages/environmental-engineering-history.aspx.
Campbell, D. T. (1974). "Downward causation in hierarchically organised biological systems". *Studies in the philosophy of biology*, 179–186.
Catalano, G. D. (2022). Beyond traditional engineering: Green, humanitarian, social justice, and omnium approaches. In D. P. Michelfelder & N. Doorn (Eds.), *The Routledge Handbook of the Philosophy of Engineering.*
Catalano G. D. (2014). Engineering ethics: Peace, justice, and the earth. *Synthesis Lectures on Engineers, Technology and Society* (2nd ed.).
Christie, P. (2020). *Science Writer for Global Change.* Retrieved from Science Writer for Global Change: https://www.peterchristiesciencecommunication.com/
D. Smith, D. S. (2020). *Yellowstone Wolves: Science and Discovery in the World's First National Park.* Chicago: University of Chicago Press.
Darwin, C. (2003). *On the Origin of the Species 150th Anniversary.* New York: Signet.
Davis, M. (1998). *Thinking Like an Engineer.* Oxford University Press.
Descartes, R. (2015). *The Collected Works of Rene Descartes: The Complete Works.* New York: Pergamon Media.
Dutcher, J. (2019). *Running with Wolves: Our Story of Life with the Sawtooth Pack.* New York: National Geographic Press.
Eco, U. (1978). *A Theory of Semiotics.* Bloomington: Indiana University Press.
Einstein, A. (2014). *The World as I See It.* New York: Create Space Independent Platform.

Emery, N. (2016). *Bird Brain: An Exploration of Avian Intelligence*. Princeton, New Jersey: Princeton University Press.
Federico Martinelli, R. S. (2015). *Advanced methods of plant disease detection. A review. Agronomy for Sustainable Development,*. Springer Verlag/EDP Sciences/INRA, 35 (1), pp.1–25. https://doi.org/10.1007/s13593-014-0246-1.
Gillispie, C. C. (1997). *Pierre-Simon Laplace, 1749–1827*. Princeton: Princeton University Press.
Goethe, J. W. (2014). *Faust I & II, Volume 2: Goethe's Collected Works - Updated Edition*. Princeton: Princeton Classics.
Gribbin, J. (2019). *Six Impossible Things: The Mystery of the Quantum World*. Cambridge: The MIT Press.
Hawkins, S. (2003). *On The Shoulders Of Giants*. Running Press Adult; 1st Printing edition.
Heath, S. T. (1991). *Greek Astronomy*. Dover Press.
Johnson, L. (1989). *A Morally Deep World*. Cambridge, Mass.: Cambridge University Press.
Kevin M. Esvelt, A. L. (2019). *Concerning RNA-guided gene drives for the alteration of wild populations*. Elife.
Kim, J. (1984). "Epiphenomenal and supervenient causation". *Midwest studies in philosophy 9*, 257–270.
L.M. Lederman. (2010). *Quantum Physics for Poets*. New York: Prometheus.
L.Vaughan-Lee. (2016). *Spiritual Ecology: The Cry of the Earth*. Los Angeles, CA: The Golden Sufi Center.
Lakota Prayer. (n.d.).
Leopold, A. (1986). *A Sand County Almanac*. New York: Ballantine Press.
Loy, D. (1997). *Nonduality: A Study in Comparative Philosophy*. Humanity Books.
Manesh, S. M. (2014). *Ethical issues of transplanting organs from transgenic animals into human beings. Cell Journal (Yakhteh) 2014;16:353–353*. Cell Journal (Yakhteh) 2014;16:353–353.
Marshall, J. (2001). *Keep Going*.
McDonough, M. (2002). *Cradle to Cradle: remaking the Way We Make Things*. Northpoint Press.
Mitcham, C. (2010). *Humanitarian Engineering*. San Rafael, CA: Morgan Claypool.
Navajo Times. (2022).
Newton, S. I. (2016). *The Principia: The Authoritative Translation and Guide: Mathematical Principles of Natural Philosophy*. Berkeley: University of California Press.
NSPE. (2019). *Code of Ethics*. Retrieved from NSPE Code of Ethics: https://www.nspe.org/resources/ethics/code-ethics
Passsino, K. (2017). *Principles of Humanitarian Engineering: A Ten-Part Series of Mini Lessons in Design for Global Development*. Columbus, Ohio: https://www.engineeringforchange.org/news/humanitarian-engineering-principle-one-focus-on-people/.
Peacocke, C. (1995). *A Study of Concepts (Representation and Mind) New edition*. New York: Bradford Books.
Peitgen, H.-O., Jürgens, H., Saupe, D., Maletsky, E., Perciante, T., & and Yunker, L. (1991). *Fractals for the Classroom: Strategic Activities Volume One*. New York: Springer Verlag.
Peterson, C. (2020). *Wolves and black-billed magpies scavenge at a dump where carcasses are stored in Yellowstone National Park*. National Geogrpahic Animals News.
Popper, K. (2002). *The Logic of Scientific Discovery*. New York: Routledge.
Ptolemy, C. (2014). *The Almagest: Introduction to Mathematics of the Heavens*. New York: Green Lion Press.
Ratzinger, J. (2010). *The Ratzinger Reader: Mapping a Theological Journey Paperback*. London: T&T Clark.
Ripple, W. J., Beschta, R. L., Fortin, J. K., & Robbins, C. T. (2014). Trophic cascades from wolves to grizzly bears in Yellowstone. *Journal of Animal Ecology, 83*, 223–233.

Rodriguez, E. (2016). *Issues in genome editing using Crispr/Cas9 system.* Journal of Clinical Research and Bioethics 7:266–266.

Seattle, C. (2020). *Chief Seattle's Speech.* Retrieved from Fresh Reads: https://www.thefreshreads.com/chief-seattles-speech/

Shelton, R., & Cliffe, J. A. (1999). *"Spherical Cows".* https://en.wikipedia.org/wiki/Spherical_cow.

Steinbuch, Y. (2020, May 19). NASA scientists detect evidence of parallel universe where time runs backward. *NY Post.*

Swimme, B. (2019). *Hidden Heart of the Cosmos.* San Francisco: Orbis.

Veitch, R. D. (2020). *Selected Philosophical Writings: Discourse on Method, Meditations on First Philosophy, Selections from the Principles of Philosophy.* Independently published.

The manufacturer's authorised representative in the EU is Springer Nature Customer Service Centre GmbH, Europaplatz 3, 69115 Heidelberg, Germany. If you have any concerns regarding our products, please contact ProductSafety@springernature.com

Printed and bound by CPI Group (UK) Ltd, Croydon, CR0 4YY

25/03/2026

02078169-0017